BURLEIGH DODDS SCIENCE: INSTANT INSIGHTS

NUMBER 17

# Sensor technologies in livestock management

burleigh dodds
SCIENCE PUBLISHING

Published by Burleigh Dodds Science Publishing Limited
82 High Street, Sawston, Cambridge CB22 3HJ, UK
www.bdspublishing.com

Burleigh Dodds Science Publishing, 1518 Walnut Street, Suite 900, Philadelphia, PA 19102-3406, USA

First published 2022 by Burleigh Dodds Science Publishing Limited
© Burleigh Dodds Science Publishing, 2022. All rights reserved.

British Library Cataloguing in Publication Data
A catalogue record for this book is available from the British Library

ISBN 978-1-80146-053-8 (Print)
ISBN 978-1-80146-054-5 (ePub)

DOI: 10.19103/9781801460545

Typeset by Deanta Global Publishing Services, Dublin, Ireland

# Contents

# Series list

# Chapter 1

## Developments in on-animal sensors for monitoring livestock

*Mark Trotter, CQUniversity Institute for Future Farming Systems, Australia; Derek Bailey, New Mexico State University, USA; and Jaime Manning, Caitlin Evans, Diogo Costa, Elle Fogarty and Anita Chang, CQ University Institute for Future Farming Systems, Australia*

## 1 Introduction

Since the very beginning of animal domestication, humans have sought to understand several key attributes of the livestock under their care. Very simply, these questions can be summarised as: Where are they (location)? What are they doing or experiencing (behaviour)? Are they in a biological state that will meet my needs in terms of the purpose for which I manage them (state)?

Traditionally, these attributes have been discerned by closely watching the animals, in some cases through almost constant human association (e.g. early shepherding practices) or through observations undertaken at critical times (e.g. dairy farmers observing their cows as they are moved into the milking parlour). This resolution of monitoring, aligned with the highly trained eye of the manager, meant that any problems or challenges to the health and productivity of individual animals could be quickly identified and ameliorated.

Unfortunately, the same degree of intensity in monitoring is no longer achievable in modern livestock production systems, particularly in remote and extensive areas, as the costs of labour preclude the ability to have trained staff

http://dx.doi.org/10.19103/AS.2021.0090.01

monitoring animals at similar temporal resolutions. While much has changed in terms of the characteristics of modern animal production compared to early subsistence husbandry, modern livestock managers still seek the same information as their historic counterparts: location, behaviour and state. Given the ever-increasing pressure to reduce costs (particularly time) and the decreasing skill of farm labourers to interpret animal behaviour, producers are increasingly turning to automated remote monitoring systems to provide this key information. Sensors have been developed for a variety of purposes and come in many shapes, sizes and means of interaction with an animal. Sensors can be broadly divided into two categories: on-animal sensors (OASs) – which are attached to, or inserted into, the animal in some manner and which are the focus of this chapter – and off-animal sensors, which collect data by observing the animal without maintaining a permanent attachment. Off-animal sensors include systems that automatically weigh, collect imagery or record vocalisations of livestock. Off-animal sensors provide valuable information, and a brief discussion of the benefits of integration between the two platforms is provided in the applications section (Section 8).

One clear benefit of OASs over off-animal systems is that the former collect information wherever the animal is located. While it is true that some off-animal imaging systems can monitor animals almost constantly in confined areas (e.g. cattle feedlots and pig barns), OASs provide the unique capability of being able to monitor individual livestock in larger areas and more extensive landscapes. As a result, the animal is monitored 24 hours a day and 7 days a week. It is this feature that has proven to be one of the key strengths of these systems, as this degree of intense monitoring could never be achieved using human labour. Most livestock are essentially herbivorous prey animals, which have evolved to hide signs of pain and weakness when observed. It is this tendency for resilience that makes detecting health and welfare problems, through occasional human inspection, a challenge (McLennan, 2018). OASs provide an opportunity to overcome this challenge as they allow for changes in location and movement to be detected while the animal goes about its daily activity without interference. This provides insights that have never been available to animal managers. Remotely detecting the location of an animal also provides an additional key benefit of enabling the animal to be easily found. This simple application is valuable, especially in extensive rangeland systems, where simply finding livestock is a time-consuming and expensive process.

OASs are not the product of the digital age; they have been used, at least in a rudimentary form, for centuries. The simple cow bell collar (Fig. 1) could be described as the earliest version of a remote animal monitoring device. The gentle ringing of the bell as the animal walks enables a manager to remotely locate the animal in difficult landscapes (usually terrain or vegetation related). The sound provided by the bell also provides a clue as to the behaviour of

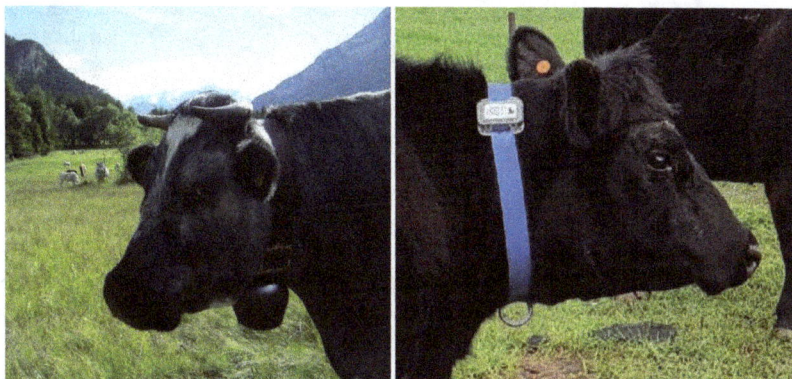

**Figure 1** The 'original' on-animal sensor, a bell on a cow in the Italian Alps North of Torino and a current commercially developed GNSS tracking collar, the Smart Paddock Blue Bell™. The original cow bell provides the livestock manager with the ability to remotely detect the location of the animal using an audible signal. The new technology enables the producer to remotely locate the cow using their smart phone. While digital technologies are transforming the livestock sector, the principle of on-animal sensing has been used for centuries (photo: Bailey and Wolchyn).

an animal; a rapidly ringing bell indicates that the cow is running, either from a predator or from the manager as they attempt to round up the herd. While the cow bell can be considered the original OASs, the digital revolution has delivered a range of new sensor platforms that have significantly advanced this concept. The development of sensors has primarily been driven by other industries, particularly the mobile phone and human wearable platforms. At first glance, it appears a simple translation to take the technology in a FitBit™ and attach it to a cow; however, in reality, there are a number of challenges that make on-animal sensing a far more complex problem.

The objective of this chapter is to provide the reader with a broad understanding of how OASs work, how they are deployed, how data from them are managed and what the data can be used for. It will also provide insights into the challenges of developing and deploying these devices in commercial livestock management operations.

## 2 Components of an on-animal sensor system

OASs systems have four broad functional components: (1) the device, which is attached to the animal and collects data; (2) the communication system, which transfers the data off the device; (3) the data handling and analytical system, which turns raw data into meaningful information; and (4) the data visualisation platform, which presents the information to the end user to enable decision making (Fig. 2). However, not all on-animal systems follow this simple linear

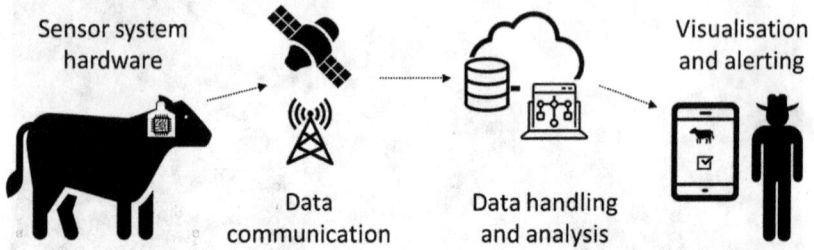

**Figure 2** Components of an on-animal sensor system. It is worth noting that not all systems follow this simple design. Many undertake data reduction or preliminary analysis on the device, some even providing alerts (e.g. a coloured light may flash to indicate a problem).

model. Some systems undertake preliminary processing or data reduction on the device while others incorporate a warning system on the device itself, such as a flashing light, to signal an issue with an individual animal. These design specifications are dictated by the specific application and environment in which the system is deployed.

The on-animal device usually consists of several key components: the sensor technology that detects the desired attribute (common sensors include accelerometers, Global Navigation Satellite Systems (GNSS) and proximity loggers); a power source that may include either an energy storage system or energy harvesting system or more commonly both; a transmitting and potentially receiving radio to communicate the data with a receiver; data storage and/or processing capability; and the packaging that contains all the elements, protects them from the external environment and attaches the entire system to the animal.

## 3 Form factor and deployment mode

Key challenges in the deployment of OASs include attaching the device to the animal (i.e. form factor) and keeping it attached (i.e. device retention). For some species this is relatively easy, and for others the development of an industry-acceptable form factor remains a considerable challenge. There have been numerous options explored over the years, from the traditional collar-borne sensor systems to more novel subcutaneous implantable sensors. A full discussion of the various deployment modes follows.

### 3.1 Leg sensors

Sensors have been deployed on the legs of dairy cattle for over 40 years, initially as a research tool (Kiddy, 1977) and subsequently as commercially available systems. The earliest leg deployed sensors were simple pedometers, and although modern leg-borne systems mostly use accelerometer technology they provide similar data as these original devices. The deployment of leg sensors on commercial dairy cows is made relatively easy by several key features in how these animals are managed, including the regular movement of animals through the milking parlour, safe access to the leg of an animal by the manager (often at working height), familiarity of animals with being handled around this area (as milking apparatus are put on and taken off regularly) and the generally quiet temperament of dairy cattle. Similar accessibility and temperament features of horses have enabled their use in a research context (DuBois et al., 2015). However, the deployment of leg-borne sensors on other large livestock species (e.g. beef cattle) is more problematic in a commercial context as animals are not as well habituated to humans in handling, although they have been used for research purposes (Ungar et al., 2018). The deployment of leg-borne sensors on smaller production animals (sheep and goats) has been reported in a research context (Champion et al., 1997; Barwick et al., 2018b); there are no examples of commercial systems using this mode of deployment.

## 3.2 Collar sensors

Collars provide a relatively simple attachment mode and, as discussed previously, have been used by producers for centuries with bells attached. One of the key benefits of a collar sensor is that the animal can support a much larger device weight compared to other deployment modes (particularly when compared to ear tags). Wildlife researchers suggest a maximum collar mass of no more than 5% of the animal's body weight (Cuthill, 1991), which for most production animals equates to a large potential collar weight. In practice, most collar-based livestock monitoring systems weigh much less than this, and research suggests that this deployment mode has little impact on the natural behaviour of grazing cattle (Manning et al., 2017a,b) and sheep (Hulbert et al., 1998). There are numerous collar-borne sensors currently available on the market, with most of them being targeted at the dairy industry for the detection of oestrus (e.g. Allflex SCR, Cow Monitor and Moocall). Most of these systems involve an accelerometer sensor (Kamphuis et al., 2012), although the integration of audio sensing has been evaluated (Reith and Hoy, 2012). At present, collars remain the only viable means of deploying virtual fencing systems that are currently being developed by numerous commercial entities.

## 3.3 Headstall or halter sensors

The headstall or halter is a slightly more complex version of a collar-based attachment enabling sensors to be placed in locations that are not accessible by other means, particularly over the nose and under the jaw. This allows for more specific head movements to be captured, including bite rate or rumination. While these systems have been used in a research context (Rahman et al., 2018), they are not common in commercial animal production systems (Benaissa et al., 2019). The headstall or halter sensors are the only widely commercialised applications that have been developed for monitoring horses with the primary aim of detecting parturition (e.g. Smart Halter).

### 3.4 Ear tag sensors

Since researchers first began using OASs there has been significant interest in the development of an ear tag-based form factor. This has been based on the perceived likelihood that industry acceptance will be greater using this type of device, and although there is no published evidence for this, it is a reasonable assumption (Schleppe et al., 2009; Barwick et al., 2018b). Most livestock managers are familiar with the application of ear tags as they have been used for individual animal identification purposes for decades. In some countries (e.g. Australia and New Zealand) electronic Radio Frequency Identification Device (RFID) ear tags are used as part of regulated animal traceability schemes. In addition, the application of ear tags to large livestock is a relatively safe process compared to collars and leg tags, which require animals to be restrained and/ or for the manager to work closely around the animal. While there is still some risk to the handler, the application of an ear tag is generally considered safer than other options.

One of the key challenges with ear tag monitoring devices is the weight limit that is imposed by the combination of the small pin attachment and the relatively soft tissue of the animal's ear. This area of research remains one that has largely been unexplored in any systematic fashion, although several case studies have reported the results of various ear tag weights. Currently, most commercial ear tag sensors weigh less than 40 g; for example the Smartbow weighs 35 g (Schweinzer et al., 2019), and the CowManager Sensor weighs 32 g (Zambelis et al., 2019). However, ear tag sensors as heavy as 227 g have been deployed on cattle (Schleppe et al., 2009) but have had to be removed after a short period of time. Heavier tags likely have a higher risk of negatively impacting animal welfare as a result of reduced wound healing and/or a lower tag retention. Commercial developers have explored several solutions to this problem including dual/multiple pin attachments and modified tag designs. While the long-term goal must be to reduce the weight of the tag, these design modifications may provide a reliable solution in the short term. A key gap in the literature exists around the maximum weight that can be sustained by different livestock species in different conditions and using different pin configurations and tag designs. This information would be of value to commercial developers seeking to balance functionality against retention.

### 3.5 Tail or tail-head sensors

Tail or tail-head-mounted sensors have been developed both as research tools (Krieger et al., 2018; Miller et al., 2020) and for commercial use (e.g. Moocall). These sensors are primarily designed for short-term deployment to detect calving behaviours (Mac, 2017). While the focus of tail deployed sensors has predominantly been on dairy cattle, this form of attachment could be valuable in other large ruminants (e.g. beef cattle), as the tail is usually readily and safely accessible. The challenge with long-term use of tail-head-mounted sensors would be in finding a means of attachment that both was reliable and did not eventually restrict blood flow around the tail itself.

### 3.6 'In-animal' sensors

In addition to the range of wearable technologies (OASs) reviewed earlier, there are a number of devices designed to be deployed internally, such as within the digestive system or subcutaneously.

#### 3.6.1 Intra-ruminal devices

Rumen bolus sensors or intra-ruminal devices have been used in a research context for many years (Mottram et al., 2008). Once administered, these devices sit in the reticulum or adjacent to the junction between the rumen and the reticulum (Koltes et al., 2018). Most rumen sensors record pH (Dijkstra et al., 2020), temperature (Ipema et al., 2008) and/or more recently, movements through accelerometers (Hamilton et al., 2019). Despite being widely reported in research, the uptake of these sensors in a commercial context remains limited in comparison to other OASs (Knight, 2020a,b). Rumen bolus sensors are highly influenced by the feed and water ingested by the animal (Bewley and Schutz, 2010), and while this may be a challenge in terms of detecting some characteristics (e.g. seeking to measure core body temperature), it may provide valuable insights into other applications (e.g. understanding fodder characteristics or detecting drinking behaviours). Recent research has suggested that simple analytical processes can be developed to enable data from a rumen bolus to be related to core body temperature (Lees et al., 2019).

#### 3.6.2 Intra-vaginal implants

These sensors are placed in the reproductive tract, most commonly in the vagina, and have been primarily used for detecting body temperature (Burdick et al., 2012). This form factor provides one of the few opportunities for reliably measuring core body temperature. The challenge with the use of vaginal implants is that they are usually expelled during parturition, a regular process for most production animals. This process has been leveraged to provide alerts to this event, with several devices having been developed that transmit radio signals after expulsion to alert to calving events (Chung et al., 2020).

### 3.6.3 Tympanic sensors

Tympanic or ear canal sensors have been developed to provide data on livestock body temperature. They consist of an ear tag with a probe that extends down the ear canal with a temperature sensor attached to the end. Conceptually, this is an ideal place to source data related to the core body temperature of livestock; however, several issues have been reported, including difficulty in correctly fitting the tag and subsequent displacement of the probe out of the ear canal (McCorkell et al., 2014).

### 3.6.4 Implantable sensors

Some small sensors have been developed that can be surgically implanted (Lee et al., 2016) or more simply injected (Chung et al., 2020). Most of these sensors have been used to collect animal body temperature data; however, there is an increasing interest in their use to monitor a range of other characteristics such as heart rate and metabolites (Neethirajan, 2017). These systems hold obvious value in measuring actual body temperature, which may aid in the detection of diseases. As with other in-animal sensors, implantable sensors have the advantage of higher retention as there is no external component that can be pulled out by being caught on structures in the field. The downside of implantable sensors is the invasiveness of implantation and the need to have small devices capable of long-term deployment. There is also some concern about the technology entering the food chain, if not successfully removed during the processing of meat-producing animals.

### 3.7 Novel deployment modes

In addition to the more commonly reported device types described above, some more novel strategies have been explored for sensor deployment. Glue on patches has been tested for monitoring bull mating activity (Abell et al., 2017). This particular device could prove valuable for monitoring bulls during the breeding season as other methods (collars and ear tags) are likely to be damaged if bulls interact aggressively. Other novel deployment options include horn attachments (where the animals have horns) and hair clip attachments (where an animal has a fleece or hair coat long enough); however, there is no evidence of these being formally evaluated in the literature.

# 4 Sensors

There is an obvious and inextricable relationship between the shape and attachment of the device and the type of sensor being used to collect the data of interest. While there have been many different sensors explored for use in monitoring animals, there are a handful that are now regularly used and have undergone considerable refinement in their application to livestock. They can be very broadly categorised into the different types of data they provide: location, activity and movement, and physiological.

## 4.1 Location sensors

These sensors provide information on either the absolute or relative location of an animal. Absolute location relates to an animal's geographic position and is most commonly referenced as latitude or longitude or may also be represented as X and Y (and sometimes Z) coordinates. Relative location data provide information on an animal's position relative to features of interest (e.g. water point) or other animals.

### 4.1.1 Global Navigation Satellite Systems

The geo-location technologies that use the network of space-borne satellites to provide positional information are collectively known as Global Navigation Satellite Systems (GNSSs). There are many national and collaborative multinational versions of these technologies; the most commonly referenced is the USA-controlled Global Positioning System (GPS). GNSS uses the radio signals broadcasted by the network of satellites to geolocate the receiver in a process known as trilateration. GNSS has been widely used in livestock research (Swain et al., 2011) and is increasingly being used as part of commercial animal monitoring systems. While GNSS can provide valuable data in most terrestrial environments, one of its key limitations is its current level of power consumption, which is relatively high compared to other sensors.

### 4.1.2 Local radio signal-based positioning

Another common form of radio-based location involves the use of OASs which emit a signal which is then detected by receivers enabling the location of the animal to be calculated. In these systems, the receivers are usually stationary antennas positioned in such a way as to optimise the signal reception and enable the calculation of the animal's location through either time of flight or signal strength (Hindermann et al., 2020). There are several examples in the literature of these systems applied to both intensive animal production systems (Wolfger et al., 2017) and extensive grazing environments (Menzies et al., 2016).

### 4.1.3 Location by proximity

While some systems are designed to provide an exact and absolute estimate of location, simpler technologies have been developed that provide the position of an animal relative to another animal or landscape feature. Most of these systems have been developed using a radio signal, with the majority of modern devices employing the Bluetooth radio protocol. The sensors usually consist of both an emitter and receiver and log a 'contact' when they come into range of each other. The number of contacts logged between individual animals is considered to relate to the amount of time they spend with each other. This technology has been used to explore social relationships between individual animals for a variety of applications, from parturition detection to more complex assessments of social structure (Swain and Bishop-Hurley, 2007; Paganoni et al., 2020).

### 4.2 Activity and movement sensors

Some of the earliest sensors used in livestock research consisted of simple activity meters or vibration recording devices (Stobbs, 1970). These simple devices enabled researchers to autonomously record and understand the behavioural patterns of animals. Since this time, the ability to collect high-resolution movement and activity data has been refined considerably with the development of digital systems that can detect a range of movement types and record large amounts of high-frequency data.

### 4.2.1 Inertia Monitoring Units

Inertia Monitoring Units (IMUs) are a combination of different individual sensor systems (an accelerometer, gyroscope and magnetometer). Each sensor provides a unique piece of data that enables the calculation of an animal's movement patterns and intensity, and when used in conjunction with one another, the animal's movement path and trajectory can be determined through a process of dead reckoning. The accelerometer provides a measure of the moving force in three directions (for triaxial accelerometers with X, Y and Z axes). The gyrometer measures the sensor's angular velocity by using the earth's gravitational force to determine orientation. Finally, the magnetometer provides the direction or bearing of the sensor orientation in relation to the earth's magnetic fields.

### 4.2.2 Accelerometer

There is only limited use of IMUs within current OAS systems. Most systems simply use data from the accelerometers (without the gyroscope and magnetometer) to monitor movement and then translate this into activities of interest. Accelerometer data can be collected at a very high temporal resolution and is considered to have a relatively low power consumption. These devices are now commonly available at affordable prices and are probably the most universally applied sensor within livestock sensor systems. One of the key challenges with accelerometers is dealing with the large volume of data. Even simple off-the-shelf systems can record data at over 500 Hz (500 times a second) across three perpendicular axes (X, Y and Z). In most cases, researchers and commercial developers reduce this data either by sampling at lower frequencies, combining multiple axes into a single measure or creating features that summarise data across a defined period of time (e.g. 5 s to 5 min), referred to as an epoch.

The data from accelerometer sensors can be used to detect a range of common or aberrant behaviours. Accelerometers have been used to distinguish between grazing or feeding, standing, walking and lying behaviours (Diosdado et al., 2015; Alvarenga et al., 2016; Barwick et al., 2018b; Fogarty et al., 2020). A more subtle behaviour, in terms of the movement of the animal, is rumination, a highly valuable behaviour in terms of understanding an animal's health and welfare. Many commercial systems in the dairy industry can detect and quantify rumination with variable results in independent tests (Wolfger et al., 2015; Pereira et al., 2018). Reports of scientific instruments successfully detecting rumination in the literature have been largely isolated to headstall or collar deployments (Shen et al., 2020; Tamura et al., 2020), although it has been reported that rumen-based accelerometers have also been successful (Hamilton et al., 2019).

Probably the most widely used applications for such sensor systems in a commercial context are collars and ear tags for the detection of oestrus events to improve the timing of artificial insemination. This application relies on accelerometer sensors to detect the increased activity associated with an oestrus event in dairy cows (Roelofs and Van Erp-van der Kooij, 2018). Accelerometers have also been used to detect specific behaviours associated with unusual behavioural activities that might provide opportunities for management intervention such as parturition events (Smith et al., 2020), disease-related behaviours (Tobin et al., 2020) and heat stress events (Davison et al., 2020).

### 4.3 Physiological sensors

There are a range of sensors that have been developed to monitor the physiological characteristics of livestock. Perhaps the simplest characteristic to measure, and one of most interest to the livestock industry to date, is body temperature. There are numerous examples of internal sensors such as rumen boluses, rectal or vaginal implants being used to provide measures of body temperature (Koltes et al., 2018; Lees et al., 2019; Baida et al., 2021). There is some conjecture around the interpretation of each deployment mode's data in relation to what it actually measures, whether core body or peripheral temperature (Godyń et al., 2019). Two types of ear tag sensors have been developed for commercial use. The first involves an ear-tag-mounted sensor with a probe that is placed in the ear canal (McCorkell et al., 2014), and the second is mounted on the ear and measures the ear surface temperature (Koltes et al., 2018). While the collection of ear surface temperature is likely to be the easiest and less intrusive of all systems, the relationship between ear surface temperature and core body temperature remains poorly understood. Subcutaneous temperature sensors have been used in some animals (Koltes et al., 2018); however, their widespread application remains limited most likely due to concerns around meat contamination should the device inadvertently enter the food chain.

Rumen pH is a physiological characteristic that has been extensively explored using reticulorumen devices. The primary target for these devices has been the detection of acidosis in cattle, a metabolic disorder commonly associated with high concentrate diets. While many systems have been developed within this research domain, their extension as widespread commercial tools remains limited as the pH sensor itself has a limited lifespan within the rumen (Hamilton et al., 2019).

One other key physiological sensor type of interest is that which measures the heart rate. Several systems have been explored to provide heart rate data ranging from electrocardiography (ECG) to photoplethysmography (PPG) techniques that are used in modern fitness trackers (Nie et al., 2020). For animals in extensive grazing environments, the adaption of these systems to an on-animal sensor will likely be necessary. In comparison, intensive livestock industries may benefit from heart rate monitor systems adapted to off-animal sensor systems (e.g. cameras). Heart rate data is of significant interest to livestock researchers for a range of reasons, from monitoring welfare to understanding the energetics of grazing animals (Laister et al., 2011; Kovács et al., 2014). However, its development for commercial situations remains isolated to applications within the equine industry where temporary fitment of electrodes can be achieved.

## 4.4 Audio sensors

Audio sensing systems have been explored in the context of off-animal deployment modes primarily for the intensive animal industries. In this situation, they have been applied to detect a range of sounds associated with abnormal or disease states (Chung et al., 2013; Huang et al., 2019). The development of audio sensing in an on-animal form factor remains limited. This is likely due to higher levels of processing power required to handle high volumes of audio data and the infrequent vocalisations of interest (Bishop et al., 2019). Audio sensing could potentially enable the measurement of ingestive behaviours and feed intake at pasture (Galli et al., 2006; Sheng et al., 2020), predation detection (from alert vocalisations) and mother/offspring interactions. However, a significant body of development work remains before these can be realised in a commercial context.

### 4.5 Future sensors that warrant further investigation

There are a handful of sensor systems in development in other industries that could have applications within the livestock sector and are worth considering. Implantable devices that monitor blood chemistry and metabolites are rapidly evolving and have primarily been used in human medicine (Yang, 2018). While this technology could provide key insights into stress-related responses of animals as a means to improve animal management, the need for surgical implantation, high initial costs and challenges around energy management may limit short-term adoption in a commercial context. However, it is likely that these sensor systems will provide key insights for animal scientists in a research context.

## 5 Energy management for on-animal sensors

Powering on-animal sensor systems remains one of the key challenges for developers. The general increase in energy density of storage systems, improvements in energy harvesting technologies, along with the reduction in power utilisation of sensor systems have resulted in smaller form factor systems being achieved. However, there is still room for improvement in this area, including two key issues that need to be considered in energy management, the storage of energy and where relevant the recharging of this storage system.

### 5.1 Energy storage

Most on-animal sensor systems integrate an energy storage system. This commonly takes the form of a battery that can hold energy (in some form of chemistry) for a long period of time and gradually supply it to the sensor system. There are some sensor systems reportedly being developed using capacitors (where the energy is stored as an electric field as opposed to the chemical energy in batteries). Systems reliant on batteries are of two types: rechargeable (secondary) and non-rechargeable (primary). Many research devices (Trotter et al., 2010) and some commercial systems reportedly use primary batteries due to their high energy density, where recharging is considered unviable. Rechargeable batteries are probably the most widely used system across all current commercial and research-grade OAS systems. Within this category, the most commonly used chemistry is that based on Lithium, the lightest weight, most energy-dense and affordable option.

### 5.2 Energy harvesting

Recharging batteries or harvesting energy for the direct operation of OASs remains another key challenge for the livestock industry. A handful of different technologies have been evaluated, including kinetic motion harvesting and inductive coupling. However, the only realistic option at this stage appears to be the collection of solar radiation using photo-voltaic (PV) technology. All current commercial on-animal sensor systems that have an energy harvesting system use PV technology (e.g. CeresTag), with small solar panels (PV cells) placed on the tag or collar. The use of PV systems on ear tags remains challenging as the available area is necessarily smaller and the placement of the tag on the ear can cause problems with an orientation towards the sun. Most early ear tag sensor systems were developed with the idea of having the device facing forward (inside the ear); however, this can result in shading of the tag, which reduces energy harvest. PV energy harvest on collars appears a viable option as a larger surface area of PV cells can be deployed when compared to ear tags.

A key issue with the use of PV energy harvesting is that animals will actively seek shade during the hottest parts of the day, which compromises the amount of solar energy that can be harvested during these optimal radiation times. One alternative for energy harvesting, as previously mentioned, is inductive coupling which is thought to provide a feasible option for intensively managed animals that can be restrained in a small space (Minnaert et al., 2018).

## 6 Communication and data transfer

There are several means of retrieving the data from the on-animal sensor, ranging from simple manual downloading to real-time radio telecommunications. Manual downloading of sensors has been used for many years in the context of research and is still used today for the collection of very high-resolution data where the transfer would be problematic due to the volume of information collected. Data transfer systems can be broadly classified by the distance over which they transmit, from short-range local to longer-range terrestrial and satellite systems.

## 6.1 Short-range terrestrial radio options

There are a number of commercially available systems that use short-range radio connections, particularly the Bluetooth, Wi-Fi and Zigbee standards. The limitation with these systems is that the animal must be located within the range of the receiving antenna. This varies according to the individual platform but is generally limited to distances less than 150 m. While useful in some intensive livestock industries, these systems pose limitations for use in extensively grazed livestock. Despite this, many systems still use short-range terrestrial radio and rely either on the animal returning to a common point (e.g. water trough) or as part of a network of connected nodes distributed across the landscape. The latter is known as a wireless sensor network and consists of a series of nodes that initially collect and then 'hop' the data using a series of short-range receivers/transmitters (Huircán et al., 2010). Short-range radio systems are likely to remain valuable options for the intensive livestock industries as the lower power consumption and reduced infrastructure costs make them more feasible in these situations.

## 6.2 Longer-range terrestrial radio options

For animals in extensive grazing systems, longer-range radio connectivity is considered more ideal. Generally, these systems work up to a range of 10–20 km, but this varies considerably depending on the specific radio system and the terrain in which it is deployed. There are a number of radio protocols that have been used in this context, but the most common is the Low-power Long-range Wide Area Network (LoRaWAN, commonly shortened to LoRa). LoRa radio connectivity has been used in many of the current commercially available on-animal sensor systems (e.g. Moovement, Smart Paddock and Cattle Watch); however, the objective assessment of these systems remains lacking in the literature. Other similar radio protocols such as the proprietary Sigfox platform have also been used for communications, but with less wide application.

## 6.3 Satellite-based radio options

While satellite communication has been frequently used in a research context, its application in commercial systems has been limited. Some systems use satellites to transfer data to the cloud once it is collected from the field (using a LoRa system). Recent trials with direct to low earth orbit satellite communication from an ear tag have been successful (Bishop-Hurley, 2021) and are currently used in at least one commercial system (CeresTag).

## 6.4 Hybrid communication systems

Some systems have been proposed and developed which integrate different components of various communication platforms to form an optimal solution.

Perhaps the most common is the use of short-range communication ear tags (deployed on many animals) with a single animal carrying a collar that acts as a mobile gateway. This system allows for the deployment of a large number of low-cost and low-energy usage devices with short-range data transfer, combined with a small number of higher-cost, higher-energy usage devices that handle data transfer over a long range.

# 7 Data management, reduction and analysis

The management of data and analysis is dealt with in detail elsewhere in this book; however, it is worthwhile discussing the relationship between sensor type, energy management, communications and the challenge of data processing and analysis. The volume of data produced by sensors varies greatly and is sometimes limited by the energy demand. For example, GNSS has relatively large power consumption and as a result many devices are programmed to collect location data sparingly. Conversely, accelerometer sensors have a relatively low power usage and, consequently, are frequently run continuously and at a sub-second rate. In terms of data production, GNSS consequently produces only a small amount of data while the accelerometer produces large volumes. For those sensors that produce large volumes of data, a common strategy is to reduce it by summarising the data over a period of time. In the case of accelerometers, a number of 'features' are often calculated. These features range from simple summations of the three axes over time (e.g. signal magnitude area) to more complex algorithms (e.g. energy and entropy). This on-device algorithm computation is commonly referred to as edge computing (Hu et al., 2016), and the summarised output can be more efficiently communicated as the volume of data is reduced compared to the original raw data.

# 8 Applications of on-animal sensors

Although other chapters in this book deal with specific applications of OASs, it is worthwhile exploring and summarising the broad range of uses reported within the various livestock industries. When OASs were first applied, it was primarily undertaken in the context of research with the purpose of providing data to answer scientific questions. As the focus shifted to application on-farm, many new applications of OASs have been both reported and proposed.

## 8.1 Applications in the intensive small animal industries (pigs and poultry)

For the intensive industries, particularly those where animals are raised in concentrated housed systems, the potential for OASs might not be as great as in other industries. Given the number of animals per unit area, other monitoring systems, particularly off-animal systems (cameras, image analysis and audio sensing), may be a more efficient means of collecting the required information.

Accelerometer sensors have been used to classify a range of pig behaviours including parturition-related activities (Thompson et al., 2019). While collars have been used to detect different behaviours (Escalante et al., 2013), their practical use outside of research remains questionable. Applications in the pig industry have focussed on disease and welfare detection (Chapa et al., 2020). Although reports of on-animal sensors deployed in commercial pig systems can be found in the literature (e.g. ear tag, Chapa et al., 2020), there are no known industry-ready systems available for this species.

The use of OASs in chicken, turkey, duck and other avian species has been limited to more novel research applications such as using small RFID-based monitoring systems to detect nest use, feeding and range use (Li et al., 2020). Much of the sensor-based research in this sector has used off-animal sensors with a particular focus on image analysis and audio monitoring (Rowe et al., 2019).

## 8.2 Applications in the intensive large animal industries (housed dairy and beef feedlots)

The use of OASs in intensive large animal industries represents the most mature commercial market of on-animal sensor technology. There are several key drivers of this: first, the individual animals are often higher in value, making it more economically feasible; second, the data transfer requires less power as the reading antennas can be placed close to the animals. This means a smaller device can be developed as large power sources are not required.

Intensive dairy producers currently use OASs, including ear tag, collar and leg-borne sensors to detect oestrus and improve the timing of artificial insemination (Roelofs and Van Erp-van der Kooij, 2018). Maltz (2020) provides a good review and discussion of the success of sensors used for oestrus detection in the dairy industry. Other applications in the dairy industry include the detection of diseases such as mastitis, lameness, metritis and metabolic disorders (Rutten et al., 2013). The current challenge is that these issues are generally not exclusively detected and identified but rather a general alert to aberrant behaviour is provided, requiring investigation by the manager to determine a final diagnosis (Eckelkamp, 2019). Further discussion around the use of OASs in the dairy industry can be found in Michie et al. (2020) and Knight (2020a,b).

While the use of OASs in intensive beef feedlots is less developed, several studies have demonstrated the ability of OASs to detect key health issues. One of the main targets has been the detection of Bovine Respiratory Disease

(BRD). Although the results of studies using sensors to detect BRD have been mixed (McCorkell et al., 2014), there is a growing belief that this disease can be effectively detected using OASs (Marchesini et al., 2018). As of yet, there has not been widespread uptake of sensors in this industry, possibly due to the cost of the systems available.

## 8.3 On-animal sensors in the extensive animal industries

The development of OASs in large-scale extensive grazing systems is in its infancy. One of the key limitations is the challenge of achieving low power radio communication across the large landscapes common in grazing production systems. These large landscapes often come with other challenges, including variation in topography and environment. Despite these challenges, there are many systems currently in development or in the early stages of commercialisation.

There have been numerous applications proposed for OASs in the grazing sector in both research and commercial contexts. It is worth noting that there are differences between the applications developed for research purposes and those being developed for use by producers in the day-to-day management of their livestock. This often causes confusion amongst commercial developers and producers as the promotion of research applications appears irrelevant to commercial livestock managers.

### 8.3.1 Cattle

In the context of research, there have been ample studies using OASs in extensive beef cattle research (Swain et al., 2011; Bailey et al., 2018). These have focussed on quantifying the various behaviours of grazing cattle (Guo et al., 2009; Hamilton et al., 2019; Sprinkle et al., 2020) and then using this data to answer broader research questions around: landscape utilisation (Ganskopp and Bohnert, 2009), the impact of imposed treatments (Bailey et al., 2008), effect of pasture availability (Manning et al., 2017a,b), behavioural changes caused by naturally occurring events (Laporte et al., 2010) or even complex relationships between genetics and animal behaviour (Bailey et al., 2015).

The application of sensors as a management tool for producers has been less well explored. There are examples of specific applications being developed to monitor some key animal behaviours that would prove useful for bull-mounting activity (Abell et al., 2017), infectious disease detection (Tobin et al., 2020), drinking behaviour (Williams et al., 2020) and feeding behaviour as animals become nutritionally challenged (Roberts et al., 2015). However, producer surveys (Trotter et al., 2018) suggest that there may be many more ways in which these systems could bring value to the beef sector including the location of animals to improve the efficiency of mustering, detection of calving, prevention of stock theft, predation detection, monitoring for plant toxicity issues, assisting in feed supplementation decisions and genetic matching of cows and calves.

### 8.3.2 Sheep and goats

The use of OASs for smaller ruminants has largely focussed on research applications with the development of commercial systems for sheep and goats limited. Like cattle, there are numerous reports of sensors being used to classify the normal behaviours (e.g. grazing, lying, standing, walking and ruminating) of small ruminants (Umstatter et al., 2008; Giovanetti et al., 2017; Fogarty et al., 2018; Mansbridge et al., 2018; Sakai et al., 2019). More complex research applications have included the exploration of energy expenditure while grazing (Animut et al., 2005), parent and offspring interactions (Broster et al., 2012), spatial landscape utilisation (Freire et al., 2012; Akasbi et al., 2013) and response to climatic challenges (Thomas et al., 2008). Further insights into the development of sensor systems for small ruminants can be found in the review by Caja et al. (2020).

In the context of developing sensor systems for producers to use, there has been a number of studies exploring key issues such as the detection of lambing events (Dobos et al., 2014), infectious disease detection (Falzon et al., 2013), oestrus detection (Fogarty et al., 2015), predation detection (Manning et al., 2014), identification of lameness (Barwick et al., 2018a) and prediction of intake at pasture (Giovanetti et al., 2020). The survey undertaken by Trotter et al. (2018) indicated a range of other applications useful to sheep producers including general welfare monitoring, detection of stock theft, genetic matching of ewes and lambs, timing of grazing rotations and ensuring the animals have access to drinking water. As these technologies become commercially viable, it is almost certain that producers will find a range of other applications and uses that will bring value to extensively grazed sheep and goats.

## 9 Future trends

At a technical level, there are several innovations in the development of specific sensors and the way in which data are managed that will likely bring significant value to both researchers and ultimately the industry as OASs evolve.

The development of sensors that focus on more specific and measurable behaviours, physiological characteristics and/or metabolites will obviously continue to leverage the development in human-related research. Some of the emerging technologies of particular interest include the direct measurement of the heart rate, which has many applications from energetics research for improved production efficiency (Brosh et al., 2006) to stress and welfare assessment (Nie et al., 2020). Another key suite of sensors that will continue to evolve are those capable of directly monitoring the metabolic state of animals either directly or by inference. While implantable technologies remain a challenge for the livestock industries because of potential food chain contamination, the value of the data provided by these systems will keep them in consideration in the future.

One of the key future trends of livestock sensing systems will be the integration of data from multiple sensor systems to provide the key information being sought by both researchers and producers. There are some early examples of different types of on-animal sensor data being integrated including multiple site deployments of accelerometers (Thompson et al., 2019), integration of GNSS and magnetometer sensors (Guo et al., 2009), GNSS with accelerometer data (Brennan et al., 2021) and the integration of location data and accelerometer sensors for dairy cattle (Benaissa et al., 2020). In terms of detecting and diagnosing key behaviours or events in livestock, the integration of data from OASs with data from off-animal sensors will likely expand in the future. Some early examples of this research include where: OASs have been integrated with a range of milk characteristics, feed and water intake and body weight to detect mastitis and lameness in dairy cattle (Post et al., 2020); and GNSS, accelerometer and weather data were integrated to detect lambing events in sheep (Fogarty et al., 2021).

As the technical development of OASs will undoubtedly continue, their use will also expand. There are several key applications that are emerging drivers of this sector. The use of these devices is being explored by regulatory agencies seeking to monitor the location, behaviour and state of livestock. There are several government agencies already using RFID systems to monitor the movement of animals for biosecurity and market integrity. The potential to improve the functionality of these systems to more complex systems including GNSS and/or accelerometer is obvious; however, the path to realising widespread adoption remains challenging. Another key driver of this technology is product authentication. There are now numerous livestock products sold under a marketing claim ranging from 'grass-fed' to enhanced welfare management. On-animal sensor systems provide an opportunity to validate these market claims, particularly for those relating to animal welfare (Fogarty et al., 2019; Chapa et al., 2020; Maroto Molina et al., 2020; Manning et al., 2021) and are likely to see significant research and development into the future.

As OASs transform from research tools into readily available and affordable systems for livestock producers, there is likely to be a significant increase in the adoption of these systems. Already, there is a widespread utilisation in the intensive and high-value animal industries, particularly dairy. As these systems become cheaper and able to work across larger and more varied landscapes, they will almost certainly become more attractive to producers in all animal industries, firstly for beef cattle and then smaller ruminants. However, the timelines for likely adoption will always be limited by the technical feasibility of systems and the benefits brought to the producer. The economic benefits of OASs remain largely unexplored. Systems may have several applications improving financial benefits through a number of cost savings and/or productivity increases, making the economic analysis of benefits complex. In addition, there are many non-financial benefits of these systems. Numerous producers, having been exposed to the concept of OASs and the real-time data they provide, commented that there is a significant component of increased 'peace of mind' of knowing where their animals are and that they are in a satisfactory state.

The potential for OASs to transform the livestock industries is enormous. However, the research community and commercial technology developers need to concentrate their efforts on understanding what data and information are required by commercial producers to improve farm decision making and provide the benefits that these systems promise. The industry has evolved in many ways since we first started using cow bells, and this new generation technologies are the key to further revolution.

# 10 References

Abell, K. M., Theurer, M. E., Larson, R. L., White, B. J., Hardin, D. K. and Randle, R. F. 2017. Predicting bull behavior events in a multiple-sire pasture with video analysis, accelerometers, and classification algorithms. *Computers and Electronics in Agriculture* 136:221-227.

Akasbi, Z., Oldeland, J., Dengler, J. and Finckh, M. 2013. Analysis of GPS trajectories to assess goat grazing pattern and intensity in Southern Morocco. *The Rangeland Journal* 34(4):415-427.

Alvarenga, F. A. P., Borges, I., Palkovič, L., Rodina, J., Oddy, V. H. and Dobos, R. C. 2016. Using a three-axis accelerometer to identify and classify sheep behaviour at pasture. *Applied Animal Behaviour Science* 181:91-99.

Animut, G., Goetsch, A. L., Aiken, G. E., Puchala, R., Detweiler, G., Krehbiel, C. R., Merkel, R. C., Sahlu, T., Dawson, L. J., Johnson, Z. B. and Gipson, T. A. 2005. Grazing behavior and energy expenditure by sheep and goats co-grazing grass/forb pastures at three stocking rates. *Small Ruminant Research* 59(2-3):191-201.

Baida, B. E. L., Swinbourne, A. M., Barwick, J., Leu, S. T. and van Wettere, W. H. 2021. Technologies for the automated collection of heat stress data in sheep. *Animal Biotelemetry* 9(1):1-15.

Bailey, D. W., Lunt, S., Lipka, A., Thomas, M. G., Medrano, J. F., Cánovas, A., Rincon, G., Stephenson, M. B. and Jensen, D. 2015. Genetic influences on cattle grazing distribution: association of genetic markers with terrain use in cattle. *Rangeland Ecology and Management* 68(2):142-149.

Bailey, D. W., Trotter, M. G., Knight, C. W. and Thomas, M. G. 2018. Use of GPS tracking collars and accelerometers for rangeland livestock production research. *Translational Animal Science* 2(1):81–88.

Bailey, D. W., VanWagoner, H. C., Weinmeister, R. and Jensen, D. 2008. Comparison of low moisture blocks and salt for manipulating grazing patterns of beef cows. *Journal of Animal Science* 86(5):1271–1277.

Barwick, J., Lamb, D., Dobos, R., Schneider, D., Welch, M. and Trotter, M. 2018a. Predicting lameness in sheep activity using tri-axial acceleration signals. *Animals: An Open Access Journal from MDPI* 8(1):12.

Barwick, J., Lamb, D. W., Dobos, R., Welch, M. and Trotter, M. 2018b. Categorising sheep activity using a tri-axial accelerometer. *Computers and Electronics in Agriculture* 145:289–297.

Benaissa, S., Tuyttens, F. A. M., Plets, D., Cattrysse, H., Martens, L., Vandaele, L., Joseph, W. and Sonck, B. 2019. Classification of ingestive-related cow behaviours using RumiWatch halter and neck-mounted accelerometers. *Applied Animal Behaviour Science* 211:9–16.

Benaissa, S., Tuyttens, F. A. M., Plets, D., Trogh, J., Martens, L., Vandaele, L., Joseph, W. and Sonck, B. 2020. Calving and estrus detection in dairy cattle using a combination of indoor localization and accelerometer sensors. *Computers and Electronics in Agriculture* 168:105153.

Bewley, J. M. and Schutz, M. M. 2010. Recent studies using a reticular bolus system for monitoring dairy cattle core body temperature. In: *Proceedings of the First North Annuaire Conference*, Precision Dairy Management, Toronto, Canada, pp. 218–219.

Bishop, J. C., Falzon, G., Trotter, M., Kwan, P. and Meek, P. D. 2019. Livestock vocalisation classification in farm soundscapes. *Computers and Electronics in Agriculture* 162:531–542.

Bishop-Hurley, G. 2021. Preliminary test results of a sensor ear tag with satellite radio communication - CSIRO Australia. In: Trotter, M. (Ed.). Rockhampton.

Brennan, J., Johnson, P. and Olson, K. 2021. Classifying season long livestock grazing behavior with the use of a low-cost GPS and accelerometer. *Computers and Electronics in Agriculture* 181:105957.

Brosh, A., Henkin, Z., Ungar, E. D., Dolev, A., Orlov, A., Yehuda, Y. and Aharoni, Y. 2006. Energy cost of cows' grazing activity: Use of the heart rate method and the Global Positioning System for direct field estimation. *Journal of Animal Science* 84(7):1951–1967.

Broster, J. C., Rathbone, D. P., Robertson, S. M., King, B. J. and Friend, M. A. 2012. Ewe movement and ewe-lamb contact levels in shelter are greater at higher stocking rates. *Animal Production Science* 52(7):502–506.

Burdick, N. C., Carroll, J. A., Dailey, J. W., Randel, R. D., Falkenberg, S. M. and Schmidt, T. B. 2012. Development of a self-contained, indwelling vaginal temperature probe for use in cattle research. *Journal of Thermal Biology* 37(4):339–343.

Caja, G., Castro-Costa, A., Salama, A. A. K., Oliver, J., Baratta, M., Ferrer, C. and Knight, C. H. 2020. Sensing solutions for improving the performance, health and wellbeing of small ruminants. *Journal of Dairy Research* 87(S1):34–46.

Champion, R. A., Rutter, S. M. and Penning, P. D. 1997. An automatic system to monitor lying, standing and walking behaviour of grazing animals. *Applied Animal Behaviour Science* 54(4):291–305. doi: 10.1016/S0168-1591(96)01210-5.

Chapa, J. M., Maschat, K., Iwersen, M., Baumgartner, J. and Drillich, M. 2020. Accelerometer systems as tools for health and welfare assessment in cattle and pigs-A review. *Behavioural Processes* 181:104262.

Chung, H., Li, J., Kim, Y., Van Os, J. M. C., Brounts, S. H. and Choi, C. Y. 2020. Using implantable biosensors and wearable scanners to monitor dairy cattle's core body temperature in real-time. *Computers and Electronics in Agriculture* 174:105453.

Chung, Y., Oh, S., Lee, J., Park, D., Chang, H. H. and Kim, S. 2013. Automatic detection and recognition of pig wasting diseases using sound data in audio surveillance systems. *Sensors* 13(10):12929-12942.

Cuthill, I. 1991. Field experiments in animal behaviour: Methods and ethics. *Animal Behaviour* 42(6):1007-1014.

Davison, C., Michie, C., Hamilton, A., Tachtatzis, C., Andonovic, I. and Gilroy, M. 2020. Detecting heat stress in dairy cattle using neck-mounted activity collars. *Agriculture* 10(6):210.

Dijkstra, J., Van Gastelen, S., Dieho, K., Nichols, K. and Bannink, A. 2020. Review: Rumen sensors: Data and interpretation for key rumen metabolic processes. *Animal: An International Journal of Animal Bioscience* 14(S1):s176-s186.

Diosdado, J. A. V., Barker, Z. E., Hodges, H. R., Amory, J. R., Croft, D. P., Bell, N. J. and Codling, E. A. 2015. Classification of behaviour in housed dairy cows using an accelerometer-based activity monitoring system. *Animal Biotelemetry* 3(1):15.

Dobos, R. C., Dickson, S., Bailey, D. W. and Trotter, M. G. 2014. The use of GNSS technology to identify lambing behaviour in pregnant grazing Merino ewes. *Animal Production Science* 54(10):1722-1727.

DuBois, C., Zakrajsek, E., Haley, D. B. and Merkies, K. 2015. Validation of triaxial accelerometers to measure the lying behaviour of adult domestic horses. *Animal: An International Journal of Animal Bioscience* 9(1):110-114. doi: 10.1017/ S175173111400247X.

Eckelkamp, E. A. 2019. Invited review: Current state of wearable precision dairy technologies in disease detection. *Applied Animal Science* 35(2):209-220.

Escalante, H. J., Rodriguez, S. V., Cordero, J., Kristensen, A. R. and Cornou, C. 2013. Sow-activity classification from acceleration patterns: A machine learning approach. *Computers and Electronics in Agriculture* 93:17-26.

Falzon, G., Schneider, D., Trotter, M. and Lamb, D. W. 2013. Correlating movement patterns of merino sheep to faecal egg counts using global positioning system tracking collars and functional data analysis. *Small Ruminant Research* 111(1):171-174.

Fogarty, E. S., Manning, J. K., Trotter, M. G., Schneider, D. A., Thomson, P. C., Bush, R. D. and Cronin, G. M. 2015. GNSS technology and its application for improved reproductive management in extensive sheep systems. *Animal Production Science* 55(10):1272-1280. doi: 10.1071/AN14032.

Fogarty, E., Swain, D., Cronin, G. and Trotter, M. 2019. A systematic review of the potential uses of on-animal sensors to monitor the welfare of sheep evaluated using the Five Domains Model as a framework. *Animal Welfare* 28(4):407-420.

Fogarty, E. S., Swain, D. L., Cronin, G. and Trotter, M. 2018. Autonomous on-animal sensors in sheep research: A systematic review. *Computers and Electronics in Agriculture* 150:245-256.

Fogarty, E. S., Swain, D. L., Cronin, G. M., Moraes, L. E., Bailey, D. W. and Trotter, M. 2021. Developing a simulated online model that integrates GNSS, accelerometer and weather data to detect parturition events in grazing sheep: A machine learning approach. *Animals: An Open Access Journal from MDPI* 11(2):303.

Fogarty, E. S., Swain, D. L., Cronin, G. M., Moraes, L. E. and Trotter, M. 2020. Behaviour classification of extensively grazed sheep using machine learning. *Computers and Electronics in Agriculture* 169:105175.

Freire, R., Swain, D. L. and Friend, M. A. 2012. Spatial distribution patterns of sheep following manipulation of feeding motivation and food availability. *Animal: An International Journal of Animal Bioscience* 6(5):846–851.

Galli, J. R., Cangiano, C. A., Demment, M. W. and Laca, E. A. 2006. Acoustic monitoring of chewing and intake of fresh and dry forages in steers. *Animal Feed Science and Technology* 128(1–2):14–30.

Ganskopp, D. C. and Bohnert, D. W. 2009. Landscape nutritional patterns and cattle distribution in rangeland pastures. *Applied Animal Behaviour Science* 116(2–4):110–119.

Giovanetti, V., Cossu, R., Molle, G., Acciaro, M., Mameli, M., Cabiddu, A., Serra, M. G., Manca, C., Rassu, S. P. G., Decandia, M. and Dimauro, C. 2020. Prediction of bite number and herbage intake by an accelerometer-based system in dairy sheep exposed to different forages during short-term grazing tests. *Computers and Electronics in Agriculture* 175:105582.

Giovanetti, V., Decandia, M., Molle, G., Acciaro, M., Mameli, M., Cabiddu, A., Cossu, R., Serra, M. G., Manca, C., Rassu, S. P. G. and Dimauro, C. 2017. Automatic classification system for grazing, ruminating and resting behaviour of dairy sheep using a tri-axial accelerometer. *Livestock Science* 196:42–48.

Godyń, D., Herbut, P. and Angrecka, S. 2019. Measurements of peripheral and deep body temperature in cattle–A review. *Journal of Thermal Biology* 79:42–49.

Guo, Y., Poulton, G., Corke, P., Bishop-Hurley, G. J., Wark, T. and Swain, D. L. 2009. Using accelerometer, high sample rate GPS and magnetometer data to develop a cattle movement and behaviour model. *Ecological Modelling* 220(17):2068–2075.

Hamilton, A. W., Davison, C., Tachtatzis, C., Andonovic, I., Michie, C., Ferguson, H. J., Somerville, L. and Jonsson, N. N. 2019. Identification of the rumination in cattle using support vector machines with motion-sensitive bolus sensors. *Sensors* 19(5):1165.

Hindermann, P., Nüesch, S., Früh, D., Rüst, A. and Gygax, L. 2020. High precision real-time location estimates in a real-life barn environment using a commercial ultra wideband chip. *Computers and Electronics in Agriculture* 170:105250.

Hu, W., Gao, Y., Ha, K., Wang, J., Amos, B., Chen, Z., Pillai, P. and Satyanarayanan, M. 2016. Quantifying the impact of edge computing on mobile applications. In: *Proceedings of the 7th ACM SIGOPS Asia-Pacific Workshop on Systems*, p. 1–8.

Huang, J., Wang, W. and Zhang, T. 2019. Method for detecting avian influenza disease of chickens based on sound analysis. *Biosystems Engineering* 180:16–24.

Huircán, J. I., Muñoz, C., Young, H., Von Dossow, L., Bustos, J., Vivallo, G. and Toneatti, M. 2010. ZigBee-based wireless sensor network localization for cattle monitoring in grazing fields. *Computers and Electronics in Agriculture* 74(2):258–264. doi: 10.1016/j.compag.2010.08.014.

Hulbert, I. A. R., Wyllie, J. T. B., Waterhouse, A., French, J. and McNulty, D. 1998. A note on the circadian rhythm and feeding behaviour of sheep fitted with a lightweight GPS collar. *Applied Animal Behaviour Science* 60(4):359–364.

Ipema, A. H., Goense, D., Hogewerf, P. H., Houwers, H. W. J. and Van Roest, H. 2008. Pilot study to monitor body temperature of dairy cows with a rumen bolus. *Computers and Electronics in Agriculture* 64(1):49-52.

Kamphuis, C., DelaRue, B., Burke, C. R. and Jago, J. 2012. Field evaluation of 2 collar-mounted activity meters for detecting cows in estrus on a large pasture-grazed dairy farm. *Journal of Dairy Science* 95(6):3045-3056.

Kiddy, C. A. 1977. Variation in physical activity as an indication of estrus in dairy cows. *Journal of Dairy Science* 60(2):235-243.

Knight, C. H. 2020a. DairyCare 'blueprint for action': Husbandry for wellbeing. *Journal of Dairy Research* 87(S1):1-8.

Knight, C. H. 2020b. Review: Sensor techniques in ruminants: More than fitness trackers. *Animal: An International Journal of Animal Bioscience* 14(S1):s187-s195.

Koltes, J. E., Koltes, D. A., Mote, B. E., Tucker, J. and Hubbell, D. S. III. 2018. Automated collection of heat stress data in livestock: New technologies and opportunities. *Translational Animal Science* 2(3):319-323.

Kovács, L., Jurkovich, V., Bakony, M., Szenci, O., Póti, P. and Tozser, J. 2014. Welfare implication of measuring heart rate and heart rate variability in dairy cattle: Literature review and conclusions for future research. *Animal: An International Journal of Animal Bioscience* 8(2):316-330.

Krieger, S., Sattlecker, G., Kickinger, F., Auer, W., Drillich, M. and Iwersen, M. 2018. Prediction of calving in dairy cows using a tail-mounted tri-axial accelerometer: A pilot study. *Biosystems Engineering* 173:79-84.

Laister, S., Stockinger, B., Regner, A.-M., Zenger, K., Knierim, U. and Winckler, C. 2011. Social licking in dairy cattle–Effects on heart rate in performers and receivers. *Applied Animal Behaviour Science* 130(3-4):81-90.

Laporte, I., Muhly, T. B., Pitt, J. A., Alexander, M. and Musiani, M. 2010. Effects of wolves on elk and cattle behaviors: Implications for livestock production and wolf conservation. *PLoS ONE* 5(8):e11954.

Lee, Y., Bok, J. D., Lee, H. J., Lee, H. G., Kim, D., Lee, I., Kang, S. K. and Choi, Y. J. 2016. Body temperature monitoring using subcutaneously implanted thermo-loggers from holstein steers. *Asian-Australasian Journal of Animal Sciences* 29(2):299-306.

Lees, A. M., Sejian, V., Lees, J. C., Sullivan, M. L., Lisle, A. T. and Gaughan, J. B. 2019. Evaluating rumen temperature as an estimate of core body temperature in Angus feedlot cattle during summer. *International Journal of Biometeorology* 63(7):939-947.

Li, N., Ren, Z., Li, D. and Zeng, L. 2020. Review: Automated techniques for monitoring the behaviour and welfare of broilers and laying hens: Towards the goal of precision livestock farming. *Animal: An International Journal of Animal Bioscience* 14(3):617-625.

Mac, S. 2017. *Evaluating the Ability to Detect Calving Time in Dairy Cattle Using a Precision Technology That Monitors Tail Movement.* Available at: https://digitalcommons .murraystate.edu/postersatthecapitol/2018/UK/5/,

Maltz, E. 2020. Individual dairy cow management: Achievements, obstacles and prospects. *Journal of Dairy Research* 87(2):145-157.

Manning, J., Cronin, G., González, L., Hall, E., Merchant, A. and Ingram, L. 2017a. The behavioural responses of beef cattle (*Bos taurus*) to declining pasture availability and the use of GNSS technology to determine grazing preference. *Agriculture* 7(5):45.

Manning, J., Power, D. and Cosby, A. 2021. Legal complexities of animal welfare in Australia: do on-animal sensors offer a future option? *Animals: An Open Access Journal from MDPI* 11(1):91.

Manning, J. K., Cronin, G. M., González, L. A., Hall, E. J. S., Merchant, A. and Ingram, L. J. 2017b. The effects of global navigation satellite system (GNSS) collars on cattle (Bos taurus) behaviour. *Applied Animal Behaviour Science* 187:54–59.

Manning, J. K., Fogarty, E. S., Trotter, M. G., Schneider, D. A., Thomson, P. C., Bush, R. D. and Cronin, G. M. 2014. A pilot study into the use of global navigation satellite system technology to quantify the behavioural responses of sheep during simulated dog predation events. *Animal Production Science* 54(10):1676–1681. doi: 10.1071/AN14221.

Mansbridge, N., Mitsch, J., Bollard, N., Ellis, K., Miguel-Pacheco, G. G., Dottorini, T. and Kaler, J. 2018. Feature selection and comparison of machine learning algorithms in classification of grazing and rumination behaviour in sheep. *Sensors* 18(10):3532.

Marchesini, G., Mottaran, D., Contiero, B., Schiavon, E., Segato, S., Garbin, E., Tenti, S. and Andrighetto, I. 2018. Use of rumination and activity data as health status and performance indicators in beef cattle during the early fattening period. *Veterinary Journal* 231:41–47.

Maroto Molina, F., Pérez Marín, C. C., Molina Moreno, L., Agüera Buendía, E. I. and Pérez Marín, D. C. 2020. Welfare Quality® for dairy cows: Towards a sensor-based assessment. *Journal of Dairy Research* 87(S1):28–33.

McCorkell, R., Wynne-Edwards, K., Windeyer, C., Schaefer, A. and UCVM Class of 2013 2014. Limited efficacy of Fever Tag® temperature sensing ear tags in calves with naturally occurring bovine respiratory disease or induced bovine viral diarrhea virus infection. *The Canadian Veterinary Journal* 55(7):688–690.

McLennan, K. M. 2018. Why pain is still a welfare issue for farm animals, and how facial expression could be the answer. *Agriculture* 8(8):127.

Menzies, D., Patison, K. P., Fox, D. R. and Swain, D. L. 2016. A scoping study to assess the precision of an automated radiolocation animal tracking system. *Computers and Electronics in Agriculture* 124:175–183.

Michie, C., Andonovic, I., Davison, C., Hamilton, A., Tachtatzis, C., Jonsson, N., Duthie, C. A., Bowen, J. and Gilroy, M. 2020. The Internet of Things enhancing animal welfare and farm operational efficiency. *Journal of Dairy Research* 87(S1):20–27.

Miller, G. A., Mitchell, M., Barker, Z. E., Giebel, K., Codling, E. A., Amory, J. R., Michie, C., Davison, C., Tachtatzis, C., Andonovic, I. and Duthie, C. A. 2020. Using animal-mounted sensor technology and machine learning to predict time-to-calving in beef and dairy cows. *Animal: An International Journal of Animal Bioscience* 14(6):1304–1312.

Minnaert, B., Thoen, B., Plets, D., Joseph, W. and Stevens, N. 2018. Wireless energy transfer by means of inductive coupling for dairy cow health monitoring. *Computers and Electronics in Agriculture* 152:101–108.

Mottram, T., Lowe, J., McGowan, M. and Phillips, N. 2008. Technical note: A wireless telemetric method of monitoring clinical acidosis in dairy cows. *Computers and Electronics in Agriculture* 64(1):45–48.

Neethirajan, S. 2017. Recent advances in wearable sensors for animal health management. *Sensing and Bio-Sensing Research* 12:15–29.

Nie, L., Berckmans, D., Wang, C. and Li, B. 2020. Is continuous heart rate monitoring of livestock a dream or is it realistic? A review. *Sensors* 20(8):2291.

Paganoni, B., Macleay, C., van Burgel, A. and Thompson, A. 2020. Proximity sensors fitted to ewes and rams during joining can indicate the birth date of lambs. *Computers and Electronics in Agriculture* 170:105249.

Pereira, G. M., Heins, B. J. and Endres, M. I. 2018. Technical note: Validation of an ear-tag accelerometer sensor to determine rumination, eating, and activity behaviors of grazing dairy cattle. *Journal of Dairy Science* 101(3):2492-2495.

Post, C., Rietz, C., Büscher, W. and Müller, U. 2020. Using sensor data to detect lameness and mastitis treatment events in dairy cows: A comparison of classification models. *Sensors* 20(14):3863.

Rahman, A., Smith, D. V., Little, B., Ingham, A. B., Greenwood, P. L. and Bishop-Hurley, G. J. 2018. Cattle behaviour classification from collar, halter, and ear tag sensors. *Information Processing in Agriculture* 5(1):124-133.

Reith, S. and Hoy, S. 2012. Relationship between daily rumination time and estrus of dairy cows. *Journal of Dairy Science* 95(11):6416-6420.

Roberts, J., Trotter, M., Schneider, D., Lamb, D., Hinch, G. and Dobos, R. 2015. Daily grazing time of free-ranging cattle as an indicator of available feed. In: *Proceedings of the 7th European Conference on Precision Livestock Farming*.

Roelofs, J. and Van Erp-van der Kooij, E. 2018. Estrus detection tools and their applicability in cattle: Recent and perspectival situation. *Animal Reproduction* 12(3):498-504.

Rowe, E., Dawkins, M. S. and Gebhardt-Henrich, S. G. 2019. A systematic review of precision livestock farming in the poultry sector: is technology focussed on improving bird welfare? *Animals: An Open Access Journal from MDPI* 9(9):614.

Rutten, C. J., Velthuis, A. G. J., Steeneveld, W. and Hogeveen, H. 2013. Invited review: sensors to support health management on dairy farms. *Journal of Dairy Science* 96(4):1928-1952.

Sakai, K., Oishi, K., Miwa, M., Kumagai, H. and Hirooka, H. 2019. Behavior classification of goats using 9-axis multi sensors: The effect of imbalanced datasets on classification performance. *Computers and Electronics in Agriculture* 166:105027.

Schleppe, J. B., Lachapelle, G., Booker, C. W. and Pittman, T. 2009. Challenges in the design of a GNSS ear tag for feedlot cattle. *Computers and Electronics in Agriculture* 70(1):84-95.

Schweinzer, V., Gusterer, E., Kanz, P., Krieger, S., Süss, D., Lidauer, L., Berger, A., Kickinger, F., Öhlschuster, M., Auer, W., Drillich, M. and Iwersen, M. 2019. Evaluation of an ear-attached accelerometer for detecting estrus events in indoor housed dairy cows. *Theriogenology* 130:19-25.

Shen, W., Cheng, F., Zhang, Y., Wei, X., Fu, Q. and Zhang, Y. 2020. Automatic recognition of ingestive-related behaviors of dairy cows based on triaxial acceleration. *Information Processing in Agriculture* 7(3):427-443.

Sheng, H., Zhang, S., Zuo, L., Duan, G., Zhang, H., Okinda, C., Shen, M., Chen, K., Lu, M. and Norton, T. 2020. Construction of sheep forage intake estimation models based on sound analysis. *Biosystems Engineering* 192:144-158.

Smith, D., McNally, J., Little, B., Ingham, A. and Schmoelzl, S. 2020. Automatic detection of parturition in pregnant ewes using a three-axis accelerometer. *Computers and Electronics in Agriculture* 173:105392.

Sprinkle, J. E., Sagers, J. K., Hall, J. B., Ellison, M. J., Yelich, J. V., Brennan, J. R., Taylor, J. B. and Lamb, J. B. 2020. Predicting cattle grazing behavior on rangeland using accelerometers. *Rangeland Ecology and Management* 76:157-170.

Stobbs, T. 1970. Automatic measurement of grazing time by dairy cows on tropical grass and legume pastures. *Tropical Grasslands* 4(3):237–244.

Swain, D. L. and Bishop-Hurley, G. J. 2007. Using contact logging devices to explore animal affiliations: Quantifying cow–calf interactions. *Applied Animal Behaviour Science* 102(1–2):1–11.

Swain, D. L., Friend, M. A., Bishop-Hurley, G. J., Handcock, R. N. and Wark, T. 2011. Tracking livestock using global positioning systems–Are we still lost? *Animal Production Science* 51(3):167–175.

Tamura, T., Chida, Y. and Okada, K. 2020. Short communication: Detection of mastication speed during rumination in cattle using 3-axis, neck-mounted accelerometers and fast Fourier transfer algorithm. *Journal of Dairy Science* 103(8):7180–7187.

Thomas, D. T., Wilmot, M. G., Alchin, M. and Masters, D. G. 2008. Preliminary indications that Merino sheep graze different areas on cooler days in the Southern Rangelands of Western Australia. *Australian Journal of Experimental Agriculture* 48(7):889–892.

Thompson, R. J., Matthews, S., Plötz, T. and Kyriazakis, I. 2019. Freedom to lie: how farrowing environment affects sow lying behaviour assessment using inertial sensors. *Computers and Electronics in Agriculture* 157:549–557.

Tobin, C., Bailey, D. W., Trotter, M. G. and O'Connor, L. 2020. Sensor based disease detection: A case study using accelerometers to recognize symptoms of Bovine Ephemeral Fever. *Computers and Electronics in Agriculture* 175:105605.

Trotter, M., Cosby, A., Manning, J., Thomson, M., Trotter, T., Graz, P., Fogarty, E., Lobb, A. and Smart, A. 2018. *Demonstrating the Value of Animal Location and Behaviour Data in the Red Meat Value Chain*. Meat and Livestock Australia Limited. Available at: https://www.mla.com.

Trotter, M. G., Lamb, D. W., Hinch, G. N. and Guppy, C. N. 2010. Global navigation satellite systems (GNSS) livestock tracking: System development and data interpretation. *Animal Production Science* 50(5):616–623.

Umstatter, C., Waterhouse, A. and Holland, J. P. 2008. An automated sensor-based method of simple behavioural classification of sheep in extensive systems. *Computers and Electronics in Agriculture* 64(1):19–26.

Ungar, E. D., Nevo, Y., Baram, H. and Arieli, A. 2018. Evaluation of the IceTag leg sensor and its derivative models to predict behaviour, using beef cattle on rangeland. *Journal of Neuroscience Methods* 300:127–137.

Williams, L. R., Moore, S. T., Bishop-Hurley, G. J. and Swain, D. L. 2020. A sensor-based solution to monitor grazing cattle drinking behaviour and water intake. *Computers and Electronics in Agriculture* 168:105141.

Wolfger, B., Jones, B. W., Orsel, K. and Bewley, J. M. 2017. Technical note: Evaluation of an ear-attached real-time location monitoring system. *Journal of Dairy Science* 100(3):2219–2224.

Wolfger, B., Timsit, E., Pajor, E. A., Cook, N., Barkema, H. W. and Orsel, K. 2015. Technical note: Accuracy of an ear tag-attached accelerometer to monitor rumination and feeding behavior in feedlot cattle. *Journal of Animal Science* 93(6):3164–3168.

Yang, G.-Z. 2018. *Implantable Sensors and Systems: From Theory to Practice*. Springer, Cham.

Zambelis, A., Wolfe, T. and Vasseur, E. 2019. Technical note: Validation of an ear-tag accelerometer to identify feeding and activity behaviors of tiestall-housed dairy cattle. *Journal of Dairy Science* 102(5):4536–4540.

# Chapter 2

## Poultry welfare monitoring: wearable technologies

*Dana L. M. Campbell, CSIRO, Australia; and Marisa A. Erasmus, Purdue University, USA*

## 1 Introduction

Globally, the trend for changes in modern poultry farming of broilers and layers is towards improved welfare. This is in response to both consumer pressures for welfare-friendly products originating from ethical farming systems, and to ensure sustainability of the industries in a changing climate with a growing population. With these goals in mind, industry trends or legislative changes have seen many countries moving towards alternative housing systems to the conventional layer cage (reviewed in Sandilands, 2019, Chapter 11) or indoor broiler shed (reviewed in de Jong, 2019, Chapter 10). These farming system changes have increased the environmental complexity in which the chickens are housed as well as amplified the number and intricacy of management decisions that must be made. For example, aviary systems for laying hens typically have multiple tiers and a litter area that require birds to be competent at moving around to avoid injury or death (Stratmann et al., 2015; Campbell et al., 2016a; Fulton, 2019). In broiler systems, for example, enrichments must be designed so that birds will be able to use them (Norring et al., 2016; Bergmann et al., 2017; Bailie et al., 2018) and provided in sufficient quantities for optimal impact (Rodriguez-Aurrekoetxea et al., 2014).

The variation within these alternative systems allows free choices by the birds for the use of distinct system areas (e.g. indoors, outdoors, litter, structural

http://dx.doi.org/10.19103/AS.2020.0078.06

tiers), which can result in sub-populations of birds with different behavioural patterns, consequently, leading to varying welfare states. For example, free-range housing systems for both broilers and layers have indoor and outdoor areas that each have their own environmental conditions that impact the birds' welfare. Not all birds will access the outdoor areas, and those that do, spend variable amounts of time outside (Rodriguez-Aurrekoetxea et al., 2014; Campbell et al., 2017a; Larsen et al., 2017; Taylor et al., 2017a), which changes depending on daily weather and/or season (Richards et al., 2011; Taylor et al., 2017b, 2018). This individual variation can sometimes be reflected in bird health (Rodriguez-Aurrekoetxea and Estevez, 2016; Campbell et al., 2017b; Larsen et al., 2018; Taylor et al., 2018), but, to date, the full management implications of these subpopulations, and best strategies to optimise the flock overall, are not well understood. Finally, loose-housing and large flock sizes result in high numbers of chickens in contact with each other, which can increase the risk of detrimental social behaviours such as feather pecking, cannibalism, or smothering (Sherwin et al., 2010; Barrett et al., 2014). Increased environmental complexity and poorer environmental control can also decrease hygiene and elevate disease risk (Jones et al., 2015; Courtice et al., 2018).

All of these systems are aimed at improving chicken welfare, but to measure and validate this for either customer assurance or continued improvement of the systems, assessing the state of the chickens at both the flock and the individual level is necessary. Animal behaviour is one of the most important animal-based measures of animal well-being. Animals use behaviour as a coping mechanism to survive and first respond to threats and stressors in their environment (Mench, 1998). Animal behaviour is an important tool for identifying animals that are unwell and is often the first sign of poor welfare that animal caretakers notice. However, the use of visual behavioural observations is subjective, error-prone and often unreliable (Weary et al., 2009). In contrast, using precision technology to quantify behavioural changes may provide a more reliable method for identifying sickness and other risks to animal welfare. Additionally, the sensitivity of precision technology may open up the possibility of detecting more subtle changes in behaviour that could be less resilient to poor health, such as play, exploration, or grooming, and be the first indicators of disease onset (e.g. mice: Litten et al., 2008; dairy cows: Mandel et al., 2017). The application of the combination of existing and new technologies to poultry farming makes it possible to identify changes in behaviour early, non-invasively and objectively, and identify specific types of animal welfare risks.

Given the vast numbers of birds that are typically housed within a farm, the application of modern precision technology to enable automated behavioural and welfare assessment is a rapidly growing sector. Precision technology is utilised across all major livestock industries with many sophisticated on-animal devices available in the commercial market to assist with monitoring individual ruminant animal welfare, such as collars, ankle straps, ear-tags, and implantable electronics. Poultry systems represent a more unique challenge because of the sheer number of birds to be monitored, their small size, and the economic value of the individual chicken. Technology for monitoring poultry welfare can be based at the flock level, such as the detection of disease or leg health through video-based monitoring (Colles et al., 2016; Dawkins, 2019, Chapter 7; Silvera et al., 2017; Fernández et al., 2018). Technology can also be based right at the level of the individual bird using wearable electronics. The current development status and availability of technology for monitoring poultry welfare was reviewed by Sassi and colleagues (2016), with the conclusion that there are multiple potential forms of monitoring, including acoustic analysis, image analysis, environmental sensors, thermal monitoring, and movement tracking. As yet, few technologies are widely commercially used in practice. Siegford and colleagues (2016) also reviewed the role of radio signal monitoring technology to obtain individual bird data within large groups of birds. More recently, Ellen and colleagues (2019) reviewed how the use of monitoring technologies such as radio-frequency identification (RFID), sensor technologies, and computer vision could identify birds at risk for feather pecking problems and the potential for genetic selection against this. Two other review papers focus on aspects of animal health. Neethirajan (2017) argues for the integration of available sensor technologies, which include sensors to monitor animal behaviour to detect animal health risks in real time. Similarly, Astill et al. (2018) reviewed the use of various types of sensors, including biosensors and wearable sensors, specifically to detect avian influenza and concluded that the implementation of technology in poultry farming is crucial as poultry production intensifies to meet increasing demand for food.

This chapter specifically reviews the use of wearable technologies applied to chickens and how these can be used to assess and monitor chicken welfare. Discussion is included for RFID technology, accelerometers, and wearable sensors with a case study presented on how RFID technology has been used to address an industry question of optimal outdoor stocking density for free-range hens within Australia. Finally, a summary of the use of research to improve agriculture, future research directions, and sources for further information are presented.

## 2 Radio-frequency identification technology

Radio-frequency identification (RFID) technology is used across multiple livestock systems for various purposes of automated animal identification and tracking (Brown-Brandl et al., 2019; Ruiz-Garcia and Lunadei, 2011). Within poultry systems, it is a key example of a wearable technology placed on individual chickens that results in extensive automated data collection to monitor behaviour and subsequently measure and improve welfare. RFID systems operate via radio waves that communicate between an RFID tag (worn by the chicken) and a tag reader or antenna that are all connected to a host system for data collection (Roberts, 2006). Tags are typically worn on the leg (Fig. 1) but can also be attached on the wings, on backpacks, or as injectable transponders (Zaninelli et al., 2016). The systems come in different forms, falling broadly into two classes based on the type of tags that the animals wear, passive or active. Passive tags, in comparison to active tags, operate without their own battery power source and thus are smaller, cheaper, and with an unlimited lifespan, but have a shorter reading range (Ruiz-Garcia and Lunadei, 2011; Brown-Brandl et al., 2019; Ellen et al., 2019). There are also different frequencies tags can operate at (Ruiz-Garcia and Lunadei, 2011; Ellen et al., 2019), with low-frequency tags having a slower read rate compared to tags operating at high frequencies, but their signals are less impeded by water and metal. Ultra-high-frequency tags have an even faster read rate but a corresponding reduced performance around water and metals. Finally, ultra-wideband tags operate across multiple frequencies and are minimally affected by any surrounding metal. Improvements in the accuracy of tag reading and reading distance generally require trade-offs regarding tag size and power supply. RFID systems within poultry research are used to automatically track the locational movement of individual birds within a housing system and/or their use of specific resources to gain a better understanding of where chickens spend their time, what factors may lead to resource-use choices, and both individual and inter-individual behavioural patterns. One of the most frequent uses of this technology has been to examine the ranging patterns of free-range hens and broilers, but research has also looked at the automated tracking of movement patterns in aviaries, perch use, nest box use, and feeding and drinking behaviour. Examples of free-range applications of RFID technology, followed by RFID monitoring of resource use in other housing systems and what can be gained for understanding behaviour and improving welfare, are presented henceforth.

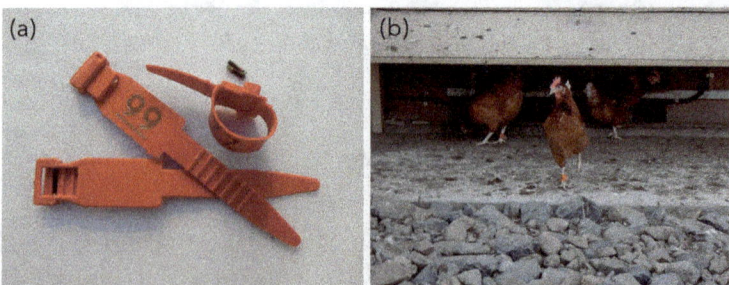

**Figure 1** (a) An adjustable leg band and a passive microchip that can be glued into the band. (b) The leg bands are worn by the hens in an experimental free-range system.

## 2.1 Radio-frequency identification technology systems and range use of chickens

To date, RFID systems using passive RFID tags (attached to wings or legs) have been applied globally to measure range usage in both laying hens and broilers, on commercial farms, and within experimental settings (Richards et al., 2011; Gebhardt-Henrich et al., 2014; Hartcher et al., 2016; Campbell et al., 2017a; Larsen et al., 2017; Taylor et al., 2017a), but only for research purposes. Free-range systems are unique from exclusively indoor-based systems as birds have a choice to go outdoors or not. Levels of range usage can vary extensively between farms, and precise range utilisation can be challenging to estimate visually because not all birds will range simultaneously (Dawkins et al., 2003; Pettersson et al., 2016). Thus, RFID technology has been increasingly used as it allows for accurate measures of when birds go outside, how long they go outside for, variables that may affect whether they range or not, and the impacts of ranging on bird health. Data gained from this chicken tracking are valuable for optimising how free-range systems are designed and managed.

The RFID systems applied to free-range farms vary, with antennas placed at the pop holes (e.g. Thurner et al., 2010; Richards et al., 2011; Campbell et al., 2017a; Taylor et al., 2017a) or at additional locations within the range area to track how far birds may move once they venture outside (e.g. Gebhardt-Henrich et al., 2014; Larsen et al., 2017). Tagged birds register on the system as they pass over an antenna. The number of antennas required is directly dependent on the size of the area to be monitored, which can sometimes restrict assessments to only a portion of a commercial facility, particularly for antennas placed on the range (Larsen et al., 2017). Antennas are typically less than 1 m in length, and read distances for wearable passive tags are generally less than 50 cm. To illustrate, an RFID system installed in an experimental free-range facility (Campbell et al., 2017a) contains decoding units which can each connect to four antennas; thus five decoders are needed to monitor nine pop holes (a total of 18 antennas, two per pop hole that are the width of a single hen only), which total approximately 50 kAUD in cost (Fig. 2). Generally, only a proportion of the total flock is tagged, allowing the estimation of flock-level patterns. Data are stored for the date/time/ID of every bird passage, often including the ability to determine the direction of travel through the use of two closely placed antennas or sensor pairs and algorithms that can filter out unpaired or 'false' readings (e.g. if a bird sits in a pop hole or if the bird does not complete an intended direction of travel).

RFID technology applied to free-range laying hen systems has been able to identify similar population-level patterns across multiple free-range settings, revealing what previous flock-level live observations could not. Not all birds will range, but the majority of hens do (67–81% ranged daily, Campbell et al.,

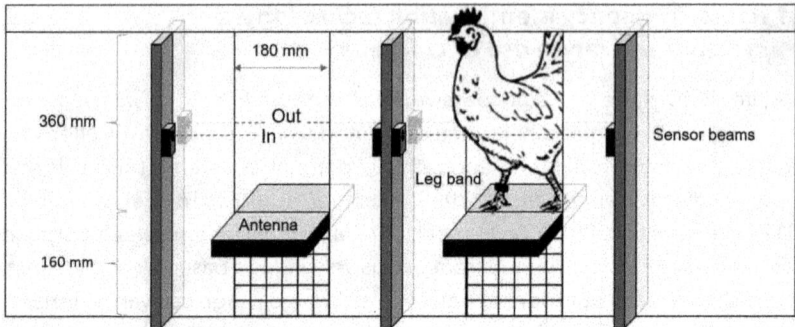

**Figure 2** The RFID system set up within the pop hole of a single pen. The antennas, sensor beams and micro-chipped leg band are indicated. Two pairs of sensor beams allow the determination of the direction of travel out of or into the pen. Reproduced from Campbell et al. (2018).

2017a; 69–82% ranged daily, Larsen et al., 2017; 80% ranged regularly, Richards et al., 2011), with individual variation in the length of ranging time and changes dependent on age, weather, and season (Pettersson et al., 2016). Of the time spent outside, less time may be spent in range areas farthest away from the shed (Larsen et al., 2017). There are fewer studies that have tracked the ranging of broilers, but work on commercial farms shows a dramatic effect of season on range usage across broiler flock replicates, where up to 33% of birds accessed the range in winter and up to 87% ranged in summer (Taylor et al., 2017a). Ranging by layers is also impacted by weather, but a single-layer flock cycle spans across multiple seasons (Richards et al., 2011).

The individual identification of chickens on free-range farms has substantially changed our understanding of bird behaviour in these systems and allowed further assessment of why this variation may exist, including the welfare implications of such variation. Behavioural tests on hens that vary in range use have shown that indoor-preferring hens are typically more fearful than hens that range daily (Campbell et al., 2016b; Hartcher et al., 2016; Larsen et al., 2018). There are also health parameters that correlate with range use, including darker combs (Larsen et al., 2018) and shorter toenails (Campbell et al., 2017b) in hens that range more or reduced range use in hens that suffer from keel bone damage (Richards et al., 2012). Broilers with greater daily range use and who range farther have reduced body weight, improved gait score, and reduced corticosterone responses to stress (Taylor et al., 2018, 2020).

Despite the wide application of RFID tracking, several questions remain. It is still unclear why some birds choose to remain indoors, if there are health, welfare, and performance implications of this across the flock or growth cycle, and if this is affected by the rearing conditions that the broilers or layers develop in. The causal relationships between ranging and welfare are also unclear; longer-term prospective studies may resolve this. The use of accelerometers may be valuable in examining the relative activity levels of birds that range or stay inside. Similarly, precisely where birds go outdoors and what they do when they are outside cannot be measured in current passive low-frequency RFID systems, but position tracking, at least, may be possible through the development of alternative tag types. Stadig and colleagues (2018a) demonstrated the application of active tags operating at ultra-wideband frequencies, worn on backpacks, which registered on seven fixed anchors located around the range site in a square formation to determine the position of individual slow-growing broilers. The system showed a mean of 68% successful registrations; future refinements would aim to improve that success rate, but, overall, the technology showed promise for measuring what passive RFID systems currently do not. However, Nakarmi et al. (2014) were able to track the locations of individuals in small group pens using passive RFID tags and a grid of antennas placed underneath the pen floor. The use of a backpack sensor versus a small leg band is potentially a greater physical strain on the chicken, but some assessments have shown minimal, short-lived differences in behaviour between tagged and untagged individuals (Stadig et al., 2018b; see Section 3).

## 2.2 Radio-frequency identification technology systems and resource use in other housing systems

RFID systems have also been applied to measure other movement patterns or resource use of both layers and broilers in other types of housing systems. Their ability to non-invasively, automatically, and continuously measure movements ensures their wide application to a range of behavioural and welfare research questions. Registration between different tag detection nodes can be compiled to determine relative activity levels of chickens within indoor housing systems, including changes across time as birds grow (Feiyang et al., 2016; Kjaer, 2017; Malchow et al., 2019). Similarly, a wireless system with backpack on-bird sensors, detection nodes, and a base station demonstrated that individual hens in a pen setting utilised the environment differently (Daigle et al., 2014). This was confirmed by Rufener and colleagues (2018), where birds equipped with active tags that were read via infrared beams in a custom-designed electronic system within an aviary housing environment exhibited patterns of movement that were consistent intra-individually, but distinct inter-individually. Classification of birds with different activity levels have consequently been used to breed high- or low-activity lines (Kjaer, 2017). Further to this, ultra-wideband tracking of where individuals go within indoor housing systems could be applied to detect behavioural differences in activity that may help identify birds susceptible to, or performing, damaging feather pecking behaviour (Rodenburg et al., 2017; Ellen et al., 2019). Correlations of movement activity with automated body weight scales and feeder sensors can also be modelled to potentially identify groups of sick, healthy, active, or inactive birds (Feiyang et al., 2016). Similarly, an ultra-high-frequency RFID system was developed by Li and colleagues (2019) to automatically detect feeding and drinking behaviour of broilers to quantify individual variation that could potentially inform management strategies. A similar system was also developed to measure feeding and nesting behaviour of laying hens in enriched cages to gain benchmark data of how hens spend their time (Li et al., 2017). This system was further used to test the impacts of varying amounts of feeder space on feeding behaviour, water intake, and egg production, which could inform both management strategies and system design refinements (Oliveira et al., 2019). A patented broiler breeder pullet precision feeding RFID system that allocated additional feed to pullets of lower body weight was demonstrated to reduce body weight CV (measure of flock uniformity) from 14% to less than 2% across the study period (Zuidhof et al., 2017). Nest boxes that automatically detect an individual hen entering can quantify nest box use, individual production rates (Oliveira et al., 2016; Zaninelli et al., 2016; Li et al., 2017; Chien and Chen, 2018), and relationships with other welfare variables such as keel fractures (Gebhardt-Henrich and Fröhlich, 2015). RFID sensors can also be placed to assess nest box type preferences (Ringgenberg et al., 2015) or preferences for different cage compartment areas (Sales et al., 2015). A system has also been applied to automatically quantify perch visits and perching duration, which could potentially be used to assess the relationships between individual perching variation and keel bone deviations or footpad dermatitis (Wang et al., 2019). The individual automated tracking of perching was able to demonstrate significant inter-pen variation despite the birds being from the same flocks, as well as significant individual variation in perch use, and showed some individuals did not perch at night despite sufficient available perching space (Wang et al., 2019). Finally, sensors on breeding cages can monitor entry of hens to facilitate the feasibility of natural mating over artificial insemination and improve animal husbandry techniques (Yu and Li, 2017).

These examples demonstrate that RFID systems are a valuable tool for research purposes where they permit the automatic quantification of movement behaviour and resource use that will provide insight into how the chickens are utilising and adapting to their housing environments. The technology could revolutionise behavioural research with precise data collected per animal relatively non-invasively. However, to date, RFID systems are not widely available for commercial implementation. The technology is not yet at a stage of being financially or logistically feasible to easily implement on farm as it is labour-intensive to install, maintain, trouble-shoot, tag birds, and monitor for lost/broken tags. Sensitive electronics in poultry barns can be challenging to maintain as the harsh environments with high dust volume and humidity degrade the equipment. Unprotected wires or system parts can also be chewed by rodents or pecked by chickens. Additionally, following system installation, there are thousands of data points being registered daily, which present challenges for compiling and interpreting that data, a point which will be discussed further in Section 6.

## 3 Wearable sensors and accelerometers

The majority of wearable motion sensors that have been investigated for poultry consist of accelerometers. Accelerometers measure the direction and magnitude of acceleration along one (single axis), two (biaxial; x and y axes), or three (triaxial; x, y and z axes) axes. Different types of accelerometers are available that can measure motion, shock, and vibration. When attached to an animal, accelerometers can measure surge, heave and sway, and provide a representation of the three-dimensional movement of the animal (reviewed in Brown et al., 2013). Advantages of accelerometers are that they are portable, wireless, and vary in price, so some relatively inexpensive accelerometers are available. However, the price is associated with the battery power and amount of memory that the device can store, which determines for how long data can be collected (Rushen et al., 2012). Therefore, unless accelerometers are used that can be paired with a receiver for continuous data downloading, which costs more and limits the size of the environment that animals can be monitored within, accelerometers have to be removed from animals to retrieve data (Rushen et al., 2012).

Accelerometers provide useful information about the animal's motion, but combining accelerometers with other types of sensors and data analysis methods can provide further context to understanding chicken behaviour. For example, the combination of temperature sensors with accelerometers enabled Okada et al. (2014) to conclude that changes in activity levels were a better early indicator of highly pathogenic avian influenza infection than a change in body temperature. Others have combined accelerometers with sensors that can track chicken location (e.g. Mica2Dot mote radio mobile sensors, Daigle et al., 2012; Tile Mate location devices, Buijs et al., 2018). Using accelerometers in conjunction with location tracking sensors may enable prediction or identification of chicken behaviour, without a need to observe chicken behaviour directly. However, the utility of combining location sensors with accelerometers has yet to be fully investigated. Instead of using combinations of sensors, another approach is to combine machine learning techniques with accelerometers to classify chicken behaviour (e.g. Abdoli et al., 2018), with the potential to track and monitor chicken behaviour automatically.

Studies that have been conducted to date indicate the potential application of wearable sensors to a wide range of behaviour and welfare scenarios. Specific applications that have been studied include the effects of wearable sensors on chicken behaviour, the ability to classify and compare behavioural activities, monitoring disease and euthanasia, and studying keel bone fractures and injuries. A discussion of each of these applications is presented in the following sections.

## 3.1 Effects of wearable sensors on chicken behaviour

Before wearable sensors can be used, they must be validated to ensure that they are not affecting the behaviour or physical ability of the animals that are under observation. Several studies have indicated that data from accelerometers can successfully be collected after a period of habituation. In addition, features of the sensor that can potentially influence bird behaviour and movement, such as sensor weight, need to be considered. As a general guide, based on similar studies in animal ecology (Murray and Fuller, 2000), Siegford et al. (2016) recommended that wearable technology for poultry should weigh less than 5% of the animal's body weight because the sensors can change bird behaviour.

The effects of wearable sensor technology on chicken behaviour and welfare have been examined in studies with laying hens (Daigle et al., 2012; Buijs et al., 2018) and one study with broiler chickens (Stadig et al., 2018b). All three studies utilised some version of a backpack to attach the sensors to the backs of the chickens. Reasons for attaching the device to the back of the chicken are that this arrangement maximises the stability of the sensor, maintains the quality of the signal, and does not result in tissue injuries to the birds (Quwaider et al., 2010). Results are generally in agreement that chickens habituate to wearing backpack sensors within a few days after sensor attachment; however, the genetic strains of chickens used, behavioural activities examined, and methods used differ across studies.

Using the sensors described in Quwaider et al. (2010), Daigle et al. (2012) examined the behaviour of commercial Hy-Line Brown laying hen pullets before and after attaching sensors at 11 weeks of age. Sensors and equipment weighed 1.07% of hens' body weight at the beginning of the study and 0.77% of hens' body weight at the end of the study. Several changes in behaviour were observed after sensors were attached. A greater percentage of hens wearing sensors used nest boxes and perches after sensor attachment than before sensors were attached. In addition, hens that wore sensors used the feeders less on 1 and 2 days after sensor attachment but used feeders more 16 days after sensors had been attached compared to hens without sensors. The percentage of agonistic interactions between hens wearing sensors and hens without sensors, where sensor-wearing hens were the aggressors, was higher than that expected at 8 and 16 days following sensor attachment. Agonistic interactions among pairs of sensor hens and pairs of non-sensor hens were no different from what would be expected. Based on their results, Daigle et al. (2012) concluded that hens habituated to the presence of the sensors after 2 weeks. In another study, Buijs et al. (2018) found no differences in aggressive behaviour or body weight when comparing sensor-wearing hens to hens that did not have sensors attached. The average body weight of birds in their study was not stated, so the weight of the sensor equipment relative to the birds' body weight is unknown. Their results revealed that there were some differences in behaviour (increased preening, sitting/lying and sidestepping/reversing, and less time standing and walking) when observed in a holding pen after being fitted with sensors than when observed in the holding pen but wearing no sensor. Drinking behaviour was also reduced after returning hens to their home pen, compared to when hens were not wearing sensors. Based on these results, Buijs et al. (2018) concluded that hens habituated to sensors relatively quickly, with few effects seen on behaviour after 2 days of wearing the sensors. Unlike the studies of Daigle et al. (2012) and Buijs et al. (2018), Stadig et al. (2018b) examined the effects of a body-mounted ultra-wideband (UWB) tag on the behaviour, leg health, and production of slow-growing male and female broiler chickens. At the time of tag attachment, the equipment weighed 1.8% of the birds' body weight. Some behavioural differences were noted, including age-related differences in how many pecks control birds directed toward birds wearing tags, and tag-wearing birds initially walked less than control birds. No differences were found between tag-wearing birds and control birds for gait, footpad dermatitis, hock dermatitis, body weight increase, or cleanliness, indicating that the tags did not adversely affect body condition and foot/leg health. Results from the aforementioned studies indicate that the attachment of wearable sensors using backpacks has some effects on chicken behaviour after sensor attachment, which illustrate the importance of providing chickens with a period of habituation before data are collected from wearable sensors. Sensor attachment does not seem to influence body weight in laying hens or broiler chickens, or footpad dermatitis and hock dermatitis of broiler chickens.

### 3.2 Classification of behavioural activities

Traditional methods of studying animal behaviour include live observations or video recordings that are then manually analysed by human observers. The use of technology to extract behavioural data and eliminate the need for labour-intensive and subjective manual analyses of video will greatly advance the study of chicken behaviour and welfare. As new technologies and data analysis methods are developed, it is becoming more feasible to extract information about animal behaviour from video and other sources, such as wearable sensors. Banerjee et al. (2012) applied machine learning to biaxial accelerometer data from laying hens to identify six distinct activities which included sitting/sleeping, standing, walking/running, feeding, drinking, and dustbathing. Accelerometers were attached to the backs of hens using a figure eight harness. In addition to the six behavioural categories, data were grouped together for higher-level classification where sitting/sleeping and standing were classified as static activities, walking/running was classified as dynamic, and feeding and drinking were classified as resource use. Data obtained from the accelerometers included time, x-axis acceleration, and y-axis acceleration; entropy and mean were extracted from the x- and y-axis acceleration. Their results indicated that machine learning techniques were able to classify behavioural activities with relatively high accuracy. More recently, Abdoli et al. (2018) developed an algorithm to detect behavioural activities from triaxial accelerometers fitted to the backs of laying hens infested with ectoparasites. Behavioural activities, including feeding/pecking, preening, and dustbathing were identified with a relatively high degree of accuracy. The authors concluded that their algorithm was able to 'learn from real, noisy and complex datasets'. These studies provide important advances in analysing chicken behaviour and have the potential to greatly improve chicken welfare once the methods become more widely available.

## 3.3 Monitoring disease and euthanasia

As poultry production increases and intensifies to meet global demands for food, risks of disease also increase. When chickens become infected with disease, especially with avian influenza, mass depopulation of flocks is conducted to prevent disease from spreading and to reduce the suffering of infected birds. The development of technologies to detect disease early and monitor disease outbreaks will be important in containing disease and safeguarding animal welfare. Furthermore, technologies that can assist with monitoring the depopulation process, especially when large numbers of chickens are killed, has the potential to ensure that depopulation is done humanely.

In order to detect avian influenza outbreaks early, Okada et al. (2009) developed a wireless, wearable sensor consisting of a triaxial accelerometer and thermistor probe (sensor to monitor chicken body temperature) to detect changes in chicken activity levels and body temperature associated with highly pathogenic avian influenza infection. Using the sensor, avian influenza infection was accurately detected several hours before death. However, not all avian influenza virus strains were associated with an increase in chicken body temperature. Consequently, Okada et al. (2014) conducted a follow-up study to investigate the feasibility of using only a triaxial accelerometer for the early detection of avian flu. Based on the differences in activity levels between pre-infection and post-infection, they were able to detect avian flu an average of 14.9 h before death occurred, which is twice as fast as what is possible using only the body temperature sensors.

Changes in broiler chicken activity levels have also been measured using accelerometers. Dawson et al. (2007, 2009) attached accelerometers to the legs of broiler chickens to (1) examine motion cessation in relation to heart relaxation recorded with an electrocardiogram (ECG) (Dawson et al., 2007) and (2) examine motion cessation in relation to brain death recorded using an electroencephalogram (EEG) (Dawson et al., 2009). Their results indicated that motion ceased before cardiac relaxation and cardiac arrest occurred following cervical dislocation (Dawson et al., 2007). Similarly, results from their other study revealed that accelerometer activity stopped before EEG activity stopped (Dawson et al., 2009). During mass depopulation with foam, it is not possible to observe chickens for signs of death. Therefore, accelerometers can provide a means of determining when the loss of brain activity occurs (Dawson et al., 2009). While it is not feasible to attach accelerometers to all chickens in a flock, attaching accelerometers to some chickens has the potential to improve disease detection and monitoring, as well as monitor mass depopulation, which can help reduce the need for people to enter the barn, further reducing the risk of spreading disease.

## 3.4 Perching, jumping, falls, and collisions

Keel bone damage and fractures are important concerns in all housing systems for laying hens, but systems that include perches are associated with greater levels of keel bone damage and fractures (reviewed in Harlander-Matauschek et al., 2015). Keel bone fractures are associated with pain (Nasr et al., 2012a) and reduced productivity (Nasr et al., 2012b). Therefore, several studies have used wearable sensors to determine factors associated with potential keel bone damage. Banerjee et al. (2014) used a backpack-mounted Mica2Dot mote radio node and triaxial accelerometer to examine when hens jumped from a perch and to calculate the hen's resulting landing force. Using their system, they were able to identify when hens had jumped from the perch, the force with which the hens landed and approximately from what height hens had jumped. Banerjee et al. (2014) suggested that the information gained from their study could be applied to other studies aimed at determining how much force results in keel bone damage and identifying particular situations that place hens at risk of keel bone fractures. More recently, LeBlanc et al. (2016) took a different approach to investigate risk factors for keel bone fractures in laying hens. In their study, triaxial accelerometers and triaxial gyroscopes were attached to hens using backpacks. Hens with various pre-existing conditions which included keel damage, footpad dermatitis, and damage of the wing feathers, as well as apparently healthy birds, were subjected to conditions that were postulated to affect their ability to maintain balance while perching. These conditions included crowding whereby 'dummy' birds were placed on either side of the hen with or without a hood placed over the head to obscure vision, and uncrowded conditions with or without the hood. Hens were tested in these conditions on a static perch and a swaying perch, operated by a motor. Hens' ability to maintain their balance while perching was affected by their physical condition; healthy hens displayed less complex movement to maintain balance while perching compared to hens with footpad dermatitis and hens with feather damage. In addition to health status, crowding negatively affected hens' balancing ability. Because hens had been habituated and trained prior to testing, LeBlanc et al. (2016) also concluded that training hens can be effective in improving their ability to maintain their balance while perching. Their study was the first to examine how both environmental factors and the health status of hens can influence their ability to balance on a perch, which provides insights into factors that may place hens at a higher risk of keel bone damage. In contrast to Banerjee et al. (2014) and LeBlanc et al. (2016) who examined factors associated with perching, Ali and Siegford (2018) did not examine perching behaviour, but used triaxial accelerometers to examine whether hens fell at night. In addition, they examined whether falling depended on whether birds were placed into aviary systems at 17 weeks compared to 25 weeks of age after being reared in floor pens, or whether being reared with access to perches compared to no perches in floor pens and placed into aviaries at 25 weeks affected falls. Birds reared in floor pens without perches and placed into aviaries at 17 weeks had more instances of falling, whereas birds reared without perches but placed into aviaries at 25 weeks had higher forces of collision and acceleration. Based on these results, Ali and Siegford (2018) concluded that accelerometers can be used to successfully detect hens' falls and collisions in aviaries at night. Through the use of wearable sensors, these studies (1) indicate that events associated with potential keel bone damage and injury can be detected non-invasively and (2) reveal important factors that can both help hens in non-cage systems, such as training (LeBlanc et al., 2016) and rearing environments (Ali and Siegford, 2018), and factors that can place hens at a greater risk of injury in non-cage systems, such as hens' physical condition (LeBlanc et al., 2016).

### 3.5 Physical activity levels

Physical activity level can be an indicator of health and welfare or can affect an animal's risk of sustaining injury. Studies have documented that chickens exhibit reduced activity levels in response to disease (e.g. Okada et al., 2014) and pain (e.g. Duncan et al., 1989), which has led to interest in using accelerometers to quantify and measure how activity levels change when chickens experience conditions expected to adversely affect their welfare. In particular, Casey-Trott and Widowski (2018) first validated that accelerometers (attached using a necklace) could detect inactivity in hens and then examined whether inactivity levels differed among hens with keel damage and hens without keel damage. Their results demonstrated that accelerometers can be used as a tool to assess chicken welfare, but contrary to what they expected, hens with keel bone damage spent less time inactive compared to healthy hens (Fig. 3).

Possible reasons for the decreased level of inactivity in hens with keel fractures are that the hens may be more restless, may spend less time standing, or may experience pain due to keel damage differently compared to how pain from other conditions would be experienced (Casey-Trott and Widowski, 2018). The accelerometers may also be useful in examining individual differences in behaviour and activity level (Casey-Trott and Widowski, 2018), which Kozak et al. (2016) suggest could be one factor affecting hens' risk of sustaining keel bone damage. Kozak et al. (2016) used accelerometers (attached using backpacks) to compare the intensity of activity level (low, moderate, and high) among laying hens of four commercial strains at three ages (10–16, 17–24, and 25–37 weeks). Their findings indicated that the accelerometers detected low-, moderate-, and high-intensity activities with a high degree of accuracy (97–98%). In addition, their results provide

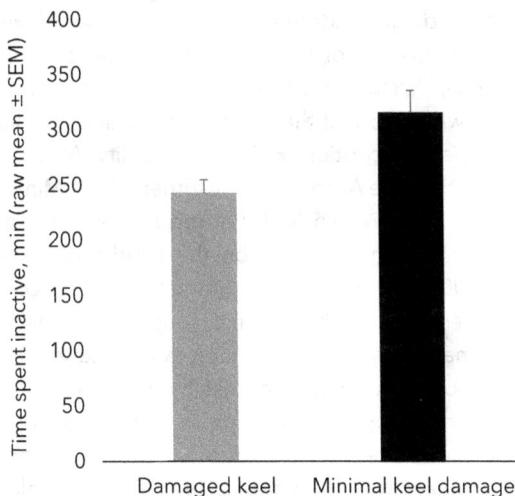

**Figure 3** The average time that hens with damaged keels and hens with little to no keel bone damage spent inactive over a 4–6 day period, determined using Actical® accelerometers. Figure produced using data from Casey-Trott and Widowski (2018). The time spent inactive was determined from activity counts obtained from the accelerometers, which differed significantly among hens with keel bone damage and hens with minimal keel bone damage.

insight into how activity levels of hens change depending on their stage of production and age, with hens entering the laying period (17–24 weeks) spending less time performing low-intensity activities and hens spending less time performing high-intensity activities as they age. Intensity level of activity differed among genetic strains as well, with white-feathered hens spending more time engaged in low-intensity activities compared to brown-feathered hens (Kozak et al., 2016). Results from Casey-Trott and Widowski (2018) and Kozak et al. (2016) indicate the potential value of using accelerometers to assess hen welfare but highlight that there may be important strain- and age-related effects that need to be considered when designing experiments and drawing conclusions about hens' activity levels.

## 4 Case study: outdoor stocking density in free-range laying hens

Within Australia, consumer preferences have driven increases in free-range laying hen systems as they are perceived to provide better hen welfare, but perceptions of how hens behave in these systems can be mismatched with reality. Free-range farms in Australia vary from mobile caravans with small flock sizes and extensive pasture access to larger-scale farms that house thousands of hens in fixed sheds with less outdoor range space per bird. Visual observations on large farms may show few birds using the range area simultaneously, which suggests not many birds go outdoors. Both these small and large-scale farms, and regardless of prevalence of hen ranging, may label their eggs as 'free-range'. Subsequently, Australian consumers felt misled by the 'free-range' label and precisely what product they were getting at a premium cost due to discrepancies between perception and system reality. As a resolution to this consumer dissatisfaction, the Australian Consumer Affairs Ministers developed an egg labelling information standard that requires egg cartons to have the hens' outdoor stocking density stated on the label and no hens are to be stocked above 10 000 hens/ha. However, at the time of this consumer debate, there was limited research on how much space hens need outdoors and how outdoor area may impact individual hen range usage. Outdoor stocking densities are based on simultaneous access by all hens (i.e. 10 000 hens/ha, gives 1 m² per hen when they are all outside together), but hens have a choice throughout the day of whether to range or not; thus, actual densities outdoors are fluid. Our experiment was designed to use RFID technology to measure individual range usage in small experimental free-range flocks with varying space outdoors as well as assess individual hen welfare and its relationship with ranging (Campbell et al., 2017a,b).

   To answer the question of the impact of outdoor stocking density on range use and hen welfare, six identical indoor pens were set up to house 150

ISA Brown hens per pen. Each pen connected to a range area of a different size to test densities of 2000, 10 000, or 20 000 hens/ha. RFID systems were placed within the pop holes, and 50% of all hens were tagged with a leg band containing a microchip. The RFID system could detect each hen that moved through the pop holes recording exact time, the direction of travel, and microchip ID. The stored data enabled the calculation of the precise duration of time each bird spent outdoors, how often they visited the range, and the length of their range visits. These hens were tracked across a period of 15 weeks, following first range access at 21 weeks of age, with periodic weighing and external welfare scoring of each hen, following a modified version of the Welfare Quality® protocol (Welfare Quality®, 2009).

The RFID tracking data showed that hens with more space outside (lowest stocking density) had longer individual visits and spent more time outside daily (32-36 weeks of age: 2000 hens/ha: 4.78 h, 10 000 hens/ha: 4.11 h, 20 000 hens/ha: 3.82 h) during ranging hours, but hens within all densities still used the range area on average for 4 h around 35 weeks of age. Across all densities there was individual variation in how often each bird went outside, including 2% of hens that were never registered as visiting the range (Campbell et al., 2017a). The birds were assessed only up to peak lay, and thus most birds were in visibly good condition, but longer ranging hours did reduce toenail length (Campbell et al., 2017b). This research showed that more area outdoors increased time spent outside, which may assist consumers in making decisions about which eggs to purchase. The individual range tracking showed there were some welfare impacts of outdoor range usage but the health benefits of ranging still require further research on more flocks. The study is an example of how wearable precision technology can be used to gather information on an industry question and provide feedback to both producers and consumers.

## 5 Conclusion

Using RFID technology and wearable sensors to analyse animal behaviour and activity levels, respectively, has several advantages, including being non-invasive, collecting quantitative information and enabling longer-term, automated monitoring of animals, at a relatively lower cost than what has previously been possible. In addition to long-term monitoring, wearable sensors enable data to be collected in real time to detect changes in animal behaviour that may be early indicators of welfare problems, with the potential to automatically issue warnings (Berckmans et al., 2015) based on these behavioural changes. Scientists envision systems that can automatically monitor and detect threats rapidly and in real time (Astill et al., 2018) to diagnose and contain the threats faster than what is currently possible. Another advantage of using technology is that it may, in some circumstances, limit the need for humans to enter the barn, which is advantageous under conditions where disease risks need to be contained.

In addition to the early identification of animal health and welfare risks, studying individual animal behaviour and behavioural characteristics can guide genetic selection strategies to improve productivity and animal welfare. With wearable sensor technologies, it is possible to collect a range of information about an individual. Sensors also make it possible to collect information from individual animals that are housed in large groups or that are housed in environments where features of the environment may impede an animal's location from being determined.

To date, the application of wearable sensor technology to measure poultry behaviour has significantly advanced our understanding of individual chickens while housed in large groups, with multiple studies highlighting how each bird may adapt and utilise the same environment in different ways. The technology has also allowed insight into impacts that chicken behaviour may have on their health, allowing strategies to be sought for improving bird welfare. If current technology continues to improve, then it could be implemented on a wider scale to inform researchers, genetic companies, and housing system design companies and enable producers to make decisions to improve their own management strategies.

## 6 Future trends in research

Wearable technologies for poultry are gaining increasing interest. Their application to date has provided a wealth of previously unattainable information, yet the technology is still in its infancy of how it could be applied for research, and commercially. As the sophistication of technology advances, smaller devices with improved battery power and operating frequencies could potentially track precisely where specific chickens are located, as well as their behavioural repertoires across large sample numbers. Thus, future research with wearable technologies will continue to have a focus at the level of the individual bird.

At the individual bird level, wearable sensors such as accelerometers provide higher resolution about certain aspects of animal behaviour than what can be observed visually; for example, accelerometers can measure changes in the number of steps, activity levels, and changes in body posture that cannot be easily detected by the human eye. However, output from accelerometers is still mainly used in conjunction with manual observations of animal behaviour because it is not yet feasible to identify specific behaviours using accelerometers alone. Future research into particular aspects of animal behaviour that can be measured and quantified using accelerometers and wearable sensors, for example, force of steps, small changes in activity levels, and distance travelled, without the need to combine accelerometer output with manual visual observations, will provide valuable information about the utility of accelerometers and other wearable sensors.

Along with the improvements in technology and interpretation of technology, wearable electronics small enough and cost effective enough for chickens that can provide real-time data on heart rate and/or body temperature could be combined with changes in behaviour to monitor for disease onset or undetected stressors. For example, these types of sensors could be used to detect changes in physiological parameters such as heart rate variability, which has been linked to stress and animal welfare (see review by Von Borell et al., 2007). Sensors that can monitor chickens' physiological as well as behavioural status may also provide insights into individual differences in animal temperament (e.g. fearfulness or flightiness) and provide new ways to investigate associations among temperament, animal welfare, and productivity. Automated tracking and monitoring of chickens across their lifetime will further elucidate the relationship between their behaviour and resulting welfare, particularly establishment of any causal relationships. While this focus on the individual will remain, it is also necessary to start looking at how that individual behaves in response to its surrounding flock mates. As more chickens are managed in loose-housed systems, their social environments correspondingly increase in complexity. Understanding the individual within their flock will help identify impacts of social facilitation on individual behaviour, the impact of an individual on the group dynamic and vice versa, or impacts of other birds on health risks of disease and injury. The establishment of subpopulations of birds within loose-housed systems may require separate management strategies such as different feed diets, to optimise the performance and welfare of all birds. With laying hens in particular, separate flocks of birds in the same environments can be vastly different in their behaviour and welfare; precise understanding of individual interactions within the flock may reveal causes for these differences and solutions to optimise every flock.

In line with this need to understand individuals within flocks, one of the major trends of future research is the necessity for a merge between animal behaviour and data science. The data collected from precision monitoring can result in millions of data points which need to be interpreted in a meaningful manner. Big data techniques such as machine learning or novel analysis and data presentation methods will be critical in maximising the knowledge gained from the data and to model both individual and flock-level patterns (e.g. Campbell et al., 2018). Researchers with data modelling skills, programming skills, and expertise in working with large datasets will be necessary for deciphering long-term precision monitoring data. Additionally, individual electronic identification of animals could contribute to the development of

Internet of Animals where real-time data on individual animals are able to be viewed online across the world (Ellen et al., 2019; Pschera, 2016).

While the focus on the individual bird is a key trend in research, the investment in the individual bird to a producer is less clear. Multiple commercial products for individual animals exist for ruminant livestock species, such as collar devices that can, for example, monitor animal location, reproductive status, or behavioural signs of illness. The economic cost and benefit of a single chicken is far less than that of a single beef cow. Thus, commercial implementations of wearable sensor technology would need to be at a comparatively minimal cost to attract producer investment. Alternatively, devices could be placed on sub-samples of birds that may be representative of the flock overall, but the exact numbers that would be required for this are still to be determined. In these cases, the need for data to be presented in a user-friendly manner is critical and likely requires the development of automated algorithms to provide simple feedback on farm. An example of application of the technology commercially could be in free-range settings to provide objective evidence of range use to consumers, or for producers to modify their management strategies accordingly. Additionally, movement pattern deviations from flock averages or individual behavioural pattern deviations could be indicative of sick individuals, or onset of disease. With increasing development of meatless protein products, animal production may become a more niche market where the consumer price will warrant a technological investment that could, for example, provide lifetime quality assurance of an individual chicken.

## 7 Where to look for further information

In addition to the literature cited in this chapter, there are several good poultry technology review articles available (Ellen et al., 2019; Sassi et al., 2016; Siegford et al., 2016; Xin and Liu, 2017). Recommendations of key societies, organisations, and websites for the latest information include the following:

- World's Poultry Science Association (www.wpsa.com) organises regular national and international poultry conferences and publishes scientific reviews in the *World's Poultry Science Journal*.
- International Society of Applied Ethology (www.applied-ethology.org) hosts both international and regional conferences annually and has several hundred members that work on welfare research including the use of poultry monitoring technologies.
- Australian Poultry Science Symposium and Poultry Science Association (www.poultryscience.org) host annual conferences presenting a wide range of poultry research.

- Poultry Hub Australia (www.poultryhub.org) based at the University of New England, Armidale, NSW, Australia, provides up-to-date scientific data on multiple aspects of poultry systems.
- The American Society of Agricultural and Biological Engineers (https://www.asabe.org/) hosts annual conferences that feature research addressing agricultural systems.

Top research centres for future research and potential collaboration include the *Animal Behaviour and Welfare Team*, CSIRO, Australia; *Center for Animal Welfare Science*, Purdue University, USA; *Department of Animals in Science and Society*, Utrecht University, The Netherlands; *Center for Animal Welfare*, University of California-Davis, USA; *Campbell Center for the Study of Animal Welfare*, University of Guelph, Canada; *Animal Behavior and Welfare Group*, Michigan State University, USA; *Department of Agricultural and Biosystems Engineering*, Iowa State University, USA; and *ZTHZ, Division of Animal Welfare*, University of Bern, Switzerland.

# 8 References

Abdoli, A., Murillo, A. C., Yeh, C.-C. M., Gerry, A. C. and Keogh, E. J. 2018. Time series classification to improve poultry welfare. *7th IEEE International Conference on Machine Learning and Applications*, pp. 635–42. doi:10.1109/ICMLA.2018.00102.

Ali, A. and Siegford, J. 2018. An approach for tracking directional activity of individual laying hens within a multi-tier cage-free housing system (aviary) using accelerometers. *Meas. Behav*, 11, p. 176–80.

Astill, J., Dara, R. A., Fraser, E. D. G. and Sharif, S. 2018. Detecting and predicting emerging disease in poultry with the implementation of new technologies and big data: a focus on avian influenza virus. *Front. Vet. Sci*. 5, 263. doi:10.3389/fvets.2018.00263.

Bailie, C. L., Baxter, M. and O'Connell, N. E. 2018. Exploring perch provision options for commercial broiler chickens. *Appl. Anim. Behav. Sci*. 200, 114–22. doi:10.1016/j.applanim.2017.12.007.

Banerjee, D., Biswas, S., Daigle, C. and Siegford, J. M. 2012. Remote activity classification of hens using wireless body mounted sensors. *Proceedings of the Ninth International Conference on Wearable and Implantable Body Sensor Networks*, 9–12 May, London, UK, pp. 107–12.

Banerjee, D., Daigle, C. L., Dong, B., Wurtz, K., Newberry, R. C., Siegford, J. M. and Biswas, S. 2014. Detection of jumping and landing force in laying hens using wireless wearable sensors. *Poult. Sci*. 93(11), 2724–33. doi:10.3382/ps.2014-04006.

Barrett, J., Rayner, A. C., Gill, R., Willings, T. H. and Bright, A. 2014. Smothering in UK free-range flocks. Part 1: incidence, location, timing and management. *Vet. Rec*. 175(1), 19. doi:10.1136/vr.102327.

Berckmans, D., Hemeryck, M., Berckmans, D., Vranken, E. and Waterschoot, T. V. 2015. Animal sound talks! Real-time sound analysis for health monitoring in livestock. In: Ni, J.-Q., Lim, T.-T. and Wang, C. (Eds), *Animal Environment and Welfare - Proceedings of the International Symposium*. China Agriculture Press, Beijing, China, pp. 215–22.

Bergmann, S., Schwarzer, A., Wilutzky, K., Louton, H., Bachmeier, J., Schmidt, P., Erhard, M. and Rauch, E. 2017. Behavior as welfare indicator for the rearing of broilers in an enriched husbandry environment – a field study. *J. Vet. Behav.* 19, 90–101. doi:10.1016/j.jveb.2017.03.003.

Brown, D. D., Kays, R., Wikelski, M., Wilson, R. and Klimley, A. P. 2013. Observing the unwatchable through acceleration logging of animal behavior. *Anim. Biotelem.* 1(1), 20. doi:10.1186/2050-3385-1-20.

Brown-Brandl, T. M., Adrio, F., Maselyne, J., Kapun, A., Hessel, E. F., Saeys, W., Van Nuffel, A. and Gallmann, E. 2019. A review of passive radio frequency identification systems for animal monitoring in livestock facilities. *Appl. Eng. Agric.* 35(4), 579–91. doi:10.13031/aea.12928.

Buijs, S., Booth, F., Richards, G., McGaughey, L., Nicol, C. J., Edgar, J. and Tarlton, J. F. 2018. Behavioural and physiological responses of laying hens to automated monitoring equipment. *Appl. Anim. Behav. Sci.* 199, 17–23. doi:10.1016/j.applanim.2017.10.017.

Campbell, D. L. M., Goodwin, S. L., Makagon, M. M., Swanson, J. C. and Siegford, J. M. 2016a. Failed landings after laying hen flight in a commercial aviary over two flock cycles. *Poult. Sci.* 95(1), 188–97. doi:10.3382/ps/pev270.

Campbell, D. L. M., Hinch, G. N., Downing, J. A. and Lee, C. 2016b. Fear and coping styles of outdoor-preferring, moderate-outdoor and indoor-preferring free-range laying hens. *Appl. Anim. Behav. Sci.* 185, 73–7. doi:10.1016/j.applanim.2016.09.004.

Campbell, D. L. M., Hinch, G. N., Dyall, T. R., Warin, L., Little, B. A. and Lee, C. 2017a. Outdoor stocking density in free-range laying hens: radio-frequency identification of impacts on range use. *Animal* 11(1), 121–30. doi:10.1017/S1751731116001154.

Campbell, D. L. M., Hinch, G. N., Downing, J. A. and Lee, C. 2017b. Outdoor stocking density in free-range laying hens: effects on behaviour and welfare. *Animal* 11(6), 1036–45. doi:10.1017/S1751731116002342.

Campbell, D. L. M., Horton, B. J. and Hinch, G. N. 2018. Using radio-frequency identification technology to measure synchronised ranging in free-range laying hens. *Animals* 8(11), 210. doi:10.3390/ani8110210.

Casey-Trott, T. M. and Widowski, T. M. 2018. Validation of an accelerometer to quantify inactivity in laying hens with or without keel-bone fractures. *Anim. Welf.* 27(2), 103–14. doi:10.7120/09627286.27.2.103.

Chien, Y. R.. and Chen, Y. X. 2018. An RFID-based smart nest box: an experimental study of laying performance and behavior of individual hens. *Sensors* 18(3), 859. doi:10.3390/s18030859.

Colles, F. M., Cain, R. J., Nickson, T., Smith, A. L., Roberts, S. J., Maiden, M. C., Lunn, D. and Dawkins, M. S. 2016. Monitoring chicken flock behaviour provides early warning of infection by human pathogen Campylobacter. *Proc. Biol. Sci.* 283(1822). doi:10.1098/rspb.2015.2323.

Courtice, J. M., Mahdi, L. K., Groves, P. J. and Kotiw, M. 2018. Spotty liver disease: a review of an ongoing challenge in commercial free-range egg production. *Vet. Microbiol.* 227, 112–8. doi:10.1016/j.vetmic.2018.08.004.

Daigle, C. L., Banerjee, D., Biswas, S. and Siegford, J. M. 2012. Noncaged laying hens remain unflappable wearing body-mounted sensors: levels of agonistic behaviors remain unchanged and resource use is not reduced after habituation. *Poult. Sci.* 91(10), 2415–23. doi:10.3382/ps.2012-02300.

Daigle, C. L., Banerjee, D., Montgomery, R. A., Biswas, S. and Siegford, J. M. 2014. Moving GIS research indoors: spatiotemperoal analysis of agricultural animals. *PLoS ONE* 9(8), e104002. doi:10.1371/journal.pone.0104002.

Dawkins, M. S., Cook, P. A., Whittingham, M. J., Mansell, K. A. and Harper, A. E. 2003. What makes free-range broiler chickens range? *In situ* measurement of habitat preference. *Anim. Behav.* 66(1), 151-60. doi:10.1006/anbe.2003.2172.

Dawson, M. D., Lombardi, M. E., Benson, E. R., Alphin, R. L. and Malone, G. W. 2007. Using accelerometers to determine the cessation of activity in broilers. *J. Appl. Poult. Res.* 16(4), 583-91. doi:10.3382/japr.2007-00023.

Dawson, M. D., Johnson, K. J., Benson, E. R., Alphin, R. L., Seta, S. and Malone, G. W. 2009. Determining cessation of brain activity during depopulation or euthanasia of broilers using accelerometers. *J. Appl. Poult. Res.* 18(2), 135-42. doi:10.3382/japr.2008-00072.

Duncan, I. J. H., Slee, G. S., Seawright, E. and Breward, J. 1989. Behavioural consequences of partial beak amputation (beak-trimming) in poultry. *Br. Poult. Sci.* 30(3), 479-88. doi:10.1080/00071668908417172.

Ellen, E. D., van der Sluis, M., Siegford, J., Guzhva, O., Toscano, M. J., Bennewitz, J., van der Zande, L. E., van der Eijk, J. A. J., de Haas, E. N., Norton, T., Piette, D., Tetens, J., de Klerk, B., Visser, B. and Rodenburg, T. B. 2019. Review of sensor technologies in animal breeding: phenotyping behaviors of laying hens to select against feather pecking. *Animals* 9(3), 108. doi:10.3390/ani9030108.

Feiyang, Z., Yueming, H., Liancheng, C., Lihong, G., Wenjie, D. and Lu, W. 2016. Monitoring behavior of poultry based on RFID radio frequency network. *Int. J. Agr. Biol. Eng.* 9, 139-47. doi:10.3965/j.ijabe.20160906.1568.

Fernández, A. P., Norton, T., Tullo, E., van Hertem, T., Youssef, A., Exadaktylos, V., Vranken, E., Guarino, M. and Berckmans, D. 2018. Real-time monitoring of broiler flock's welfare status using camera-based technology. *Biosyst. Eng.* 173, 103-14. doi:10.1016/j.biosystemseng.2018.05.008.

Fulton, R. M. 2019. Health of commercial egg laying chickens in different housing systems. *Avian Dis.* 63(3), 420-6. doi:10.1637/11942-080618-Reg.1.

Gebhardt-Henrich, S. G. and Fröhlich, E. K. F. 2015. Early onset of laying and bumblefoot favour keel bone fractures. *Animals* 5(4), 1192-206. doi:10.3390/ani5040406.

Gebhardt-Henrich, S. G., Toscano, M. J. and Fröhlich, E. K. F. 2014. Use of outdoor ranges by laying hens in different sized flocks. *Appl. Anim. Behav. Sci.* 155, 74-81. doi:10.1016/j.applanim.2014.03.010.

Harlander-Matauschek, A., Rodenburg, T. B., Sandilands, V., Tobalske, B. W. and Toscano, M. J. 2015. Causes of keel bone damage and their solutions in laying hens. *World. Poult. Sci. J.* 71(3), 461-72. doi:10.1017/S0043933915002135.

Hartcher, K. M., Hickey, K. A., Hemsworth, P. H., Cronin, G. M., Wilkinson, S. J. and Singh, M. 2016. Relationships between range access as monitored by radio frequency identification technology, fearfulness, and plumage damage in free-range laying hens. *Animal* 10(5), 847-53. doi:10.1017/S1751731115002463.

Jones, D. R., Cox, N. A., Guard, J., Fedorka-Cray, P. J., Buhr, R. J., Gast, R. K., Abdo, Z., Rigsby, L. L., Plumblee, J. R., Karcher, D. M., Robison, C. I., Blatchford, R. A. and Makagon, M. M. 2015. Microbiological impact of three commercial laying hen housing systems. *Poult. Sci.* 94(3), 544-51. doi:10.3382/ps/peu010.

Kjaer, J. B. 2017. Divergent selection on home pen locomotor activity in a chicken model: selection program, genetic parameters and direct response on activity and body weight. *PLoS ONE* 12(8), e0182103. doi:10.1371/journal.pone.0182103.

Kozak, M., Tobalske, B., Springthorpe, D., Szkotnicki, B. and Harlander-Matauschek, A. 2016. Development of physical activity levels in laying hens in three-dimensional aviaries. *Appl. Anim. Behav. Sci.* 185, 66–72. doi:10.1016/j.applanim.2016.10.004.

Larsen, H., Cronin, G. M., Gebhardt-Henrich, S. G., Smith, C. L., Hemsworth, P. H. and Rault, J. L. 2017. Individual ranging behaviour patterns in commercial free-range layers as observed through RFID tracking. *Animals* 7(3), 21. doi:10.3390/ani7030021.

Larsen, H., Hemsworth, P. H., Cronin, G. M., Gebhardt-Henrich, S. G., Smith, C. L. and Rault, J. L. 2018. Relationship between welfare and individual ranging behaviour in commercial free-range laying hens. *Animal* 12(11), 2356–64. doi:10.1017/S1751731118000022.

LeBlanc, S., Tobalske, B., Quinton, M., Springthorpe, D., Szkotnicki, B., Wuerbel, H. and Harlander-Matauschek, A. 2016. Physical health problems and environmental challenges influence balancing behaviour in laying hens. *PLoS ONE* 11(4), e0153477. doi:10.1371/journal.pone.0153477.

Li, L., Zhao, Y., Oliveira, J., Verhoijsen, W., Liu, K. and Xin, H. 2017. A UHF RFID system for studying individual feeding and nesting behaviors of group-housed laying hens. *Trans. ASABE* 60(4), 1337–47. doi:10.13031/trans.12202.

Li, G., Zhao, Y., Hailey, R., Zhang, N., Liang, Y. and Purswell, J. L. 2019. An ultra-high frequency radio frequency identification system for studying individual feeding and drinking behaviors of group-housed broilers. *Animal* 13(9), 2060–9. doi:10.1017/S1751731118003440.

Litten, K., Acevedo, A., Browne, W., Edgar, J., Mendl, M., Owen, D., Sherwin, C., Würbel, H. and Nicol, C. 2008. Towards humane end points: behavioural changes precede clinical signs of disease in a Huntington's disease model. *Proc. Biol. Sci.* 275(1645), 1865–74. doi:10.1098/rspb.2008.0388.

Malchow, J., Puppe, B., Berk, J. and Schrader, L. 2019. Effects of elevated grids on growing male chickens differing in growth performance. *Front. Vet. Sci.* 6, 203. doi:10.3389/fvets.2019.00203.

Mandel, R., Nicol, C. J., Whay, H. R. and Klement, E. 2017. Short communication: detection and monitoring of metritis in dairy cows using an automated grooming device. *J. Dairy. Sci.* 100(7), 5724–8. doi:10.3168/jds.2016-12201.

Mench, J. 1998. Why it is important to understand animal behavior. ILAR J. 39, 20–6. doi:10.1093/ilar.39.1.20.

Murray, D. L. and Fuller, M. R. 2000. A critical review of the effects of marking on the biology of vertebrates. In: Boitani, L. and Fuller, T. (Eds), *Research Techniques in Animal Ecology Controversies and Consequences*. Columbia University Press, New York, NY, pp. 15–64.

Nakarmi, A. D., Tang, L. and Xin, H. 2014. Automated tracking and behavior quantification of laying hens using 3D computer vision and radio frequency identification technologies. *Trans. ASABE* 57, 1455–72. Available at: https://lib.dr.iastate.edu/abe_eng_pubs/612.

Nasr, M. A., Nicol, C. J. and Murrell, J. C. 2012a. Do laying hens with keel bone fractures experience pain? *PLoS ONE* 7(8), e42420. doi:10.1371/journal.pone.0042420.

Nasr, M., Murrell, J., Wilkins, L. and Nicol, C. 2012b. The effect of keel fractures on egg-production parameters, mobility and behaviour in individual laying hens. *Anim. Welf.* 21(1), 127–35.doi:10.7120/096272812799129376.

Neethirajan, S. 2017. Recent advances in wearable sensors for animal health management. *Sens. Biosensing. Res.* 12, 15–29. doi:10.1016/j.sbsr.2016.11.004.

Norring, M., Kaukonen, E. and Valros, A. 2016. The use of perches and platforms by broiler chickens. *Appl. Anim. Behav. Sci.* 184, 91–6. doi:10.1016/j.applanim.2016.07.012.

Okada, H., Suzuki, K., Tsukamoto, K. and Itoh, T. 2009. Wireless sensor system for detection of avian influence outbreak farms at an early stage. *Proceedings of the IEEE Sensation*, 25–28 October, pp. 1374–7.

Okada, H., Suzuki, K., Kenji, T. and Itoh, T. 2014. Applicability of wireless activity sensor network to avian influenza monitoring system in poultry farms. *J. Sens. Technol.* 4(1), 18–23. doi:10.4236/jst.2014.41003.

Oliveira, J., Xin, H., Zhao, Y., Li, L., Liu, K. and Glaess, K. 2016. Nesting behavior and egg production pattern of laying hens in enriched colony housing. *Trans. ASABE.* doi:10.13031/aim.20162456546, 162456546.

Oliveira, J. L., Xin, H. and Wu, H. 2019. Impact of feeder space on laying hen feeding behavior and production performance in enriched colony housing. *Animal* 13(2), 374–83. doi:10.1017/S1751731118001106.

Pettersson, I. C., Freire, R. and Nicol, C. J. 2016. Factors affecting ranging behaviour in commercial free-range hens. *Worlds Poult. Sci. J.* 72(1), 137–50. doi:10.1017/S0043933915002664.

Pschera, A. 2016. *Animal Internet: Nature and the Digital Revolution* (Trans. Lauffer, E.). New Vessel Press, New York, NY.

Quwaider, M., Daigle, C., Biswas, S., Siegford, J. and Swanson, J. 2010. Development of a wireless body-mounted sensor to monitor location and activity of laying hens in a non-cage housing system. *Trans. ASABE* 53(5), 1705–13. doi:10.13031/2013.34890.

Richards, G. J., Wilkins, L. J., Knowles, T. G., Booth, F., Toscano, M. J., Nicol, C. J. and Brown, S. N. 2011. Continuous monitoring of pop hole usage by commercially housed free-range hens throughout the production cycle. *Vet. Rec.* 169(13), 338. doi:10.1136/vr.d4603.

Richards, G. J., Wilkins, L. J., Knowles, T. G., Booth, F., Toscano, M. J., Nicol, C. J. and Brown, S. N. 2012. Pop hole use by hens with different keel fracture status monitored throughout the laying period. *Vet. Rec.* 170(19), 494. doi:10.1136/vr.100489.

Ringgenberg, N., Fröhlich, E. K. F., Harlander-Matauschek, A., Toscano, M. J., Würbel, H. and Roth, B. A. 2015. Effects of variation in nest curtain design on pre-laying behaviour of domestic hens. *Appl. Anim. Behav. Sci.* 170, 34–43. doi:10.1016/j.applanim.2015.06.008.

Roberts, C. M. 2006. Radio frequency identification (RFID). *Comput. Secur.* 25(1), 18–26. doi:10.1016/j.cose.2005.12.003.

Rodenburg, T. B., Bennewitz, J., de Haas, E. N., Košťál, L., Pichová, K., Piette, D., Tetens, J., Van Der Eijk, J., Visser, B. and Ellen, E. D. 2017. The use of sensor technology and genomics to breed for laying hens that show less damaging behaviour. *8th EC-PLF*, Nantes, France, 12–14 September.

Rodriguez-Aurrekoetxea, A. and Estevez, I. 2016. Use of space and its impact on the welfare of laying hens in a commercial free-range system. *Poult. Sci.* 95(11), 2503–13. doi:10.3382/ps/pew238.

Rodriguez-Aurrekoetxea, A., Leone, E. H. and Estevez, I. 2014. Environmental complexity and use of space in slow growing free range chickens. *Appl. Anim. Behav. Sci.* 161, 86–94. doi:10.1016/j.applanim.2014.09.014.

Rufener, C., Berezowski, J., Maximiano Sousa, F., Abreu, Y., Asher, L. and Toscano, M. J. 2018. Finding hens in a haystack: consistency of movement patterns within and across individual laying hens maintained in large groups. *Sci. Rep.* 8(1), 12303. doi:10.1038/s41598-018-29962-x.

Ruiz-Garcia, L. and Lunadei, L. 2011. The role of RFID in agriculture: applications, limitations and challenges. *Comput. Electron. Agric.* 79(1), 42–50. doi:10.1016/j.compag.2011.08.010.

Rushen, J., Chapinal, N. and de Passille, A. M. 2012. Automated monitoring of behavioural-based animal welfare indicators. *Anim. Welf.* 21(3), 339–50. doi:10.7120/09627286.21.3.339.

Sales, G. T., Green, A. R., Gates, R. S., Brown-Brandl, T. M. and Eigenberg, R. A. 2015. Quantifying detection performance of a passive low-frequency RFID system in an environmental preference chamber for laying hens. *Comput. Electron. Agric.* 114, 261–8. doi:10.1016/j.compag.2015.03.008.

Sassi, N. B., Averós, X. and Estevez, I. 2016. Technology and poultry welfare. *Animals* 6(10), 62. doi:10.3390/ani6100062.

Sherwin, C. M., Richards, G. J. and Nicol, C. J. 2010. Comparison of the welfare of layer hens in 4 housing systems in the UK. *Br. Poult. Sci.* 51(4), 488–99. doi:10.1080/00071668.2010.502518.

Siegford, J. M., Berezowski, J., Biswas, S. K., Daigle, C. L., Gebhardt-Henrich, S. G., Hernandez, C. E., Thurner, S. and Toscano, M. J. 2016. Assessing activity and location of individual laying hens in large groups using modern technology. *Animals* 6(2), 10. doi:10.3390/ani6020010.

Silvera, A. M., Knowles, T. G., Butterworth, A., Berckmans, D., Vranken, E. and Blokhuis, H. J. 2017. Lameness assessment with automatic monitoring of activity in commercial broiler flocks. *Poult. Sci.* 96(7), 2013–7. doi:10.3382/ps/pex023.

Stadig, L. M., Ampe, B., Rodenburg, T. B., Reubens, B., Maselyne, J., Zhuang, S., Criel, J. and Tuyttens, F. A. M. 2018a. An automated positioning system for monitoring chickens' location: accuracy and registration success in a free-range area. *Appl. Anim. Behav. Sci.* 201, 31–9. doi:10.1016/j.applanim.2017.12.010.

Stadig, L. M., Rodenburg, T. B., Ampe, B., Reubens, B. and Tuyttens, F. A. M. 2018b. An automated positioning system for monitoring chickens' location: effects of wearing a backpack on behaviour, leg health and production. *Appl. Anim. Behav. Sci.* 198, 83–8. doi:10.1016/j.applanim.2017.09.016.

Stratmann, A., Fröhlich, E. K. F., Gebhardt-Henrich, S. G., Harlander-Matauschek, A., Würbel, H. and Toscano, M. J. 2015. Modification of aviary design reduces incidence of falls, collisions and keel bone damage in laying hens. *Appl. Anim. Behav. Sci.* 165, 112–23. doi:10.1016/j.applanim.2015.01.012.

Taylor, P. S., Hemsworth, P. H., Groves, P. J., Gebhardt-Henrich, S. G. and Rault, J. L. 2017a. Ranging behaviour of commercial free-range broiler chickens 2: individual variation. *Animals* 7(7), 55. doi:10.3390/ani7070055.

Taylor, P. S., Hemsworth, P. H., Groves, P. J., Gebhardt-Henrich, S. G. and Rault, J. L. 2017b. Ranging behaviour of commercial free-range broiler chickens 1: factors related to flock variability. *Animals* 7(7), 54. doi:10.3390/ani7070054.

Taylor, P. S., Hemsworth, P. H., Groves, P. J., Gebhardt-Henrich, S. G. and Rault, J. L. 2018. Ranging behavior relates to welfare indicators pre- and post-range access in commercial free-range broilers. *Poult. Sci.* 97(6), 1861–71. doi:10.3382/ps/pey060.

Taylor, P. S., Hemsworth, P. H., Groves, P. J., Gebhart-Henrich, S. G. and Rault, J. L. 2020. Frequent range visits further from the shed relate positively to free-range broiler chicken welfare. *Animal* 14(1), 138–49. doi:10.1017/S1751731119001514.

Thurner, S., Maier, S., Icken, W., Wendle, G. and Preisinger, R. 2010. Identification reliability of laying hens at the wide electronic pop hole. *Landtechnik* 65, 139–41.

Von Borell, E., Langbein, J., Despres, G., Hansen, S., Leterrier, C., Marchant-Forde, J., Marchant-Forde, R., Minero, M., Mohr, E., Prunier, A., Valance, D. and Veissier, I. 2007. Heart rate variability as a measure of autonomic regulation of cardiac activity for assessing stress and welfare in farm animals: a review. *Physiol. Behav.* 92(3), 293–316. doi:10.1016/j.physbeh.2007.01.007.

Wang, K., Liu, K., Xin, H., Chai, L., Wang, Y., Fei, T., Oliveira, J., Pan, J. and Ying, Y. 2019. An RFID-based automated individual perching monitoring system for group-housed poultry. Agricultural and Biosystems Engineering Publications. 1025. Available at: https://lib.dr.iastate.edu/abe_eng_pubs/1025.

Weary, D. M., Huzzey, J. M. and von Keyserlingk, M. A. G. 2009. Board-invited review: using behavior to predict and identify ill health in animals. *J. Anim. Sci.* 87, 770–7. doi.org/10.2527/jas.2008-1297

Welfare Quality®. 2009. Welfare Quality® assessment protocol for poultry (broilers, laying hens). Welfare Quality® Consortium, Lelystad, the Netherlands.

Xin, H. and Liu, K. 2017. Precision livestock farming in egg production. *Anim. Front.* 7(1), 24–31. doi:10.2527/af.2017.0105.

Yu, Y. and Li, L. 2017. RFID-based laying hen individual behavior monitoring system. *Rev. Fac. Ing.* 32, 904–9.

Zaninelli, M., Rossi, L., Costa, A., Tangorra, F. M., Guarino, M. and Savoini, G. 2016. Performance of injected RFID transponders to collect data about laying performance and behaviour of hens. *Large Anim. Rev.* 22, 77–82.

Zuidhof, M. J., Fedorak, M. V., Ouellette, C. A. and Wenger, I. I. 2017. Precision feeding: innovative management of broiler breeder feed intake and flock uniformity. *Poult. Sci.* 96(7), 2254–63. doi:10.3382/ps/pex013.

# Chapter 3

## Advances in techniques for health monitoring/disease detection in dairy cattle

*Michael Iwersen and Marc Drillich, University of Veterinary Medicine Vienna, Austria*

## 1 Introduction

In recent decades, livestock production has been characterized by the intensification and specialization of production processes leading to larger farms and, hence, to fewer farmers per animal. Thus, the animal-to-stockman ratio is increasing within the dairy sector and the available time for monitoring of an individual animal is reducing. Compared with the past, diversified family-owned farms have frequently switched to highly specialized companies that apply industrial production processes.

In order to earn a living, farmers have to house and care for a larger number of animals than in the past. Today's society, however, expects that animals are kept under high animal welfare and health standards that ensure 'a life worth living' (Webster, 2016), including that animals receive individual attention. However, in production-intensive sectors like cattle production, farmers have less time per animal as a consequence of the above-mentioned economy of scale. Within the production cycle, the farmer or stockperson has a unique role in ensuring high standards of animal welfare. Caring for livestock is, at least for the majority of farmers, a very positive experience, and positive human

http://dx.doi.org/10.19103/AS.2020.0086.04

contacts between farmer and animal have been shown to have a beneficial effect upon the welfare of both (Boivin et al., 2003). The quality of stockmanship, particularly the care and handling of animals, is responsible for some of the differences between farms and farmers in the level of animal welfare (Rushen et al., 2008). Additionally, there is evidence that good stockmanship improves productivity and is essential for an economically successful farm. Hence, reducing supervision time per animal might have negative socio-economic consequences for the stakeholders involved in the business, foremost not only for the animals and the farmer but also for their veterinarians, feed consultants, and consumers.

The observed economy of scale, as well as the growing demand for animal-derived food, will exert even stronger pressure on farm consolidation and process efficiency, inducing the creation of corporations with even larger farm units. On the other hand, these changes in farm structure offer the opportunity for the cost-effective use of sophisticated technologies, for example, automated feeding and milking systems, as well as animal health and welfare monitoring systems.

Traditional and increasingly sophisticated technologies support farmers and veterinarians in monitoring animals for the early detection of, for example, diseases or welfare issues and enable the collection of individual animal data as well as data from the entire herd and, thus, can bring the animals closer to the farmers again and assist them in their daily tasks. For instance, this includes the early and accurate detection of diseases, followed by timely veterinary intervention, which usually results in earlier recovery. Using these technologies has an impact not only on animal health and welfare but also on working routines and optimized resource management, increasing the competitiveness of the farm, in general.

## 2 Shift in the veterinary profession

With regard to animal health, the focus in the veterinary profession has shifted from the treatment of acutely diseased animals to more proactive health management of the entire herd (LeBlanc et al., 2006). This also includes ensuring high animal welfare standards. This shift in veterinary medicine has occurred in recent decades and is in accordance with changes in farmers' demands and expectations.

Currently, farmers' demand for veterinary advice goes beyond clinical activities and therapies (Pothmann et al., 2014). For example, farmers ask their veterinarians for profound support in areas of environmental protection, welfare, nutrition, grassland management, economics, and miscellaneous business management (Da Silva et al., 2006). Farmers' and veterinarians' decisions should be evidence based on the most reliable data available, preferably in real time. In this context, veterinary diagnoses contribute to the pool of farm data that can be used for herd (health) management decisions.

The above-mentioned paradigm shift in veterinary medicine and farmers' demands on ensuring animal health and productivity has brought proactive disease prevention strategies in the focus of herd health medicine. Realizing that most diseases have a multifactorial genesis led to the implementation of (farm) specific prevention programs, in particular, in early lactation. Epidemiological tools have increasingly been used to identify risk factors at the animal and herd level. As a result of this, research identified that an already subclinical form of a disease, for example, subclinical ketosis, hypocalcemia, mastitis, or endometritis, has a negative impact on individual animal health, productivity, and welfare (Walsh et al., 2007; Suthar et al., 2013; Rodriguez et al., 2017; Venjakob et al., 2017; Wagener et al., 2017; Gussmann et al., 2019) and on total herd performance. These findings led to the redefinitions of notorious and well-known diseases in a more complex and broader view (LeBlanc et al., 2006; Wagener et al., 2017).

## 2.1 Herd health management

Herd health management, also referred to as population medicine, is a multidisciplinary approach to optimize animal health, welfare, and productivity through systematic analyses involving planning, implementing, monitoring, and controlling production processes. Based on these systematic analyses, decisions are made as best and timely as possible to improve the general farm management over time (LeBlanc et al., 2006). For this, the strategies intend to gather information about the individual animal and the entire herd, striving for the final objective of assisting in decision-making processes. Typical tasks or events of a farmer and/or veterinarian, where decision-making processes are needed immediately are, for example, if and how to treat an animal with signs of diseases, or to intervene or not at parturition, but also in the case of physiological events, for example, in estrus detection.

After parturition, most larger dairies keep their fresh cows in specific pens to facilitate health monitoring and examinations (Guterbock, 2004), which is usually performed by farm personnel to identify sick cows. Not only do the training and skills but also the willingness of the personnel for fresh cow monitoring vary between farms (Espadamala et al., 2016). In a written survey of 429 farms, Heuwieser et al. (2010) reported that most dairy managers use subjective criteria to identify diseased animals, for example, general appearance (97%) and appetite (70%). Only the minority of responding participants monitor their cows by the use of objective (fever 34%) or semi-quantitative measures, for example, measuring ketones (3%) or body condition scoring (36%).

## 2.2 Epidemiology

Epidemiology deals with identifying potential causal associations between exposures and outcomes. Ultimately, research aims to 'make causal inferences about relationships between exposure and disease in the source population as a preliminary step toward developing policy and programs to maintain health and prevent disease' (Dohoo et al., 2009). From a veterinary standpoint, the exposures are often risk factors, leading to disease, or impaired animal performance as an outcome. In this context, farm-specific policies and programs consist of, for example, routine animal examinations and other prophylactic measures for disease prevention.

In recent decades, numerous large-scale observational studies have been conducted on commercial farms to identify risk factors for the development of diseases. As an example of early studies focusing on the interrelationship between animal nutrition, health, and reproductive performance of periparturient cows, Dohoo and Martin (1984), Curtis et al. (1985), and Cameron et al. (1998) should be mentioned here. These and other research groups have contributed to a profound knowledge base on the interrelationship between diseases and disorders, respectively, as well as their influence on animal performance. Additionally, Erb and Gröhn (1988), who postulated associations among metabolic disorders and risk factors [i.e. diseases and milk yield in previous lactation, dry cow nutrition, neonate factors (e.g. sex, size, survival, twins), dystocia, retained fetal membranes, and metritis], as presented in Figure 1, should also be mentioned here.

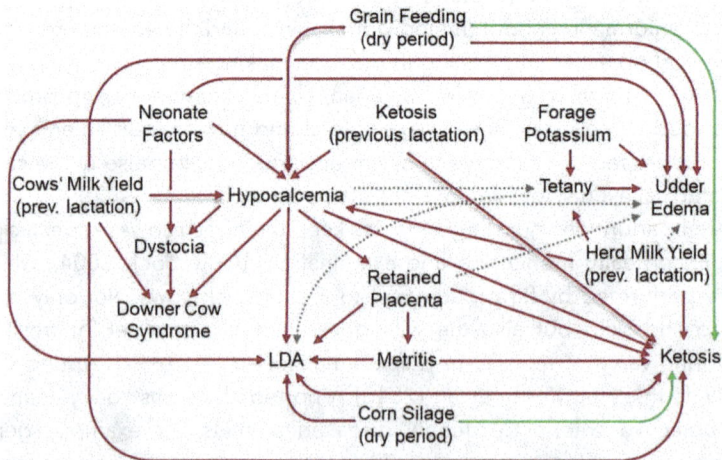

**Figure 1** Risk factors for clinical metabolic disorders in periparturient cows postulated by Erb and Gröhn (1988, modified figure). Projected effects are represented by dotted lines. Red lines indicate disease-promoting, green lines disease-inhibiting effects. LDA = left displaced abomasum.

The authors concluded that 'none of the six periparturient metabolic disorders occurs in isolation from the others' and that 'milk fever plays a central role linking to all of the other disorders' (Erb and Gröhn, 1988). Furthermore, the occurrence of ketosis was identified as a risk factor for the development of various diseases in the subsequent lactation. As ketosis can be a risk factor for and also a sequalae of the left displaced abomasum (LDA), the association between both was represented with a double-headed arrow.

Much recent research in dairy herd management focuses on the time period from late pregnancy to early lactation, which is also referred to as a 'transition period' and defined as the period from 3 weeks pre- until 3 weeks post-calving (Grummer, 1995). Dairy cows have to face many physiological changes during their transition from the dry to their lactating period. These include a temporary state of physiological immunosuppression around parturition, which, among others, is caused by a lowered bactericidal capacity of polymorphnuclear leukocytes and decreased responsiveness of lymphocytes to mitogenic agents (Detilleux et al., 1995; Nonnecke et al., 2003; Lacasse et al., 2018). Hence, animals in early lactation are at a greater risk for the occurrence of infectious diseases, such as mammary (Bradley and Green, 2004; Pyörälä, 2008) and uterine infections (Drillich and Wagener, 2018; Sheldon et al., 2020). Because of the before mentioned reasons, most health disorders occur during the transition period (Drackley, 1999; LeBlanc, 2010) and approximately 75% of all diseases take place in the first month of lactation (LeBlanc, 2006).

## 3 Information management systems

The above-mentioned findings of epidemiological research are now increasingly being applied in the daily practice of dairy farms. A fast and timely analysis is made possible by the use of information technologies, which consist of hard- and software solutions for capturing data as well as transforming them into information, which can support farmers and other stakeholders in decision making (Van Asseldonk, 1999). The active use of these systems is another cornerstone in the overall concept of early detection of problems and diseases at the herd and animal level.

The use of analog and electronic data collection systems has become a valuable tool in epidemiology as well as in herd health management. In particular, on-farm management software has been developed for data collection, integration, analyses, and easy presentation of results. While the integration of many cow-specific data, for example, on milk yield, milk solids, and veterinary diagnoses has already been possible for some decades using sophisticated software solutions, for example, DairyComp 305 (Valley Agricultural Software, Tulare, USA) and DSA Dairy-Veterinary (DSAHR,

St-Hyacinthe, Canada), the integration of external data resources, for example, on climate, feeding ration, and data exchange with other technical devices is often challenging. Hence, interfaces are needed to enable the sharing and uploading of these often heterogeneous data onto central data storage. This is one prerequisite of a more in-depth analysis of farm data and its environment in order to gather additional knowledge and enable further improvement of herd health management (Stewart et al., 1994; Wischenbart et al., 2017; Gusterer et al., 2019). Furthermore, access to a high-performance internet connection and the general availability of WLAN are further basic requirements for data exchange in real time, for linking production processes and increase our knowledge.

## 3.1 Data collection and definition of disease

In general, farmers and veterinarians do not need a huge amount of data to do their daily tasks. In this context, data collection should be as simple and timely as possible. Data collection that is too detailed may be interesting for profound in-depth analyses or from a scientific point of view, but for most dairy farms it does not seem to lead to an improvement in organizing daily tasks or in the identification and solving of problems. On the contrary, a data collection system that is too complex and that, for example, requires a detailed description or a selection from a several hundred diseases, tends to lead to frustration during data entry. This could be one reason that, quite often, only few or no data are collected on farms. Data collection should therefore be limited to a few and most common and relevant diseases occurring on farms. However, these well-defined, systematically recorded, and reliable data have to be entered consistently, in a timely manner and as easily as possible. For this further standardization of disease definition and an intuitive codification (i.e. avoiding entry of numerical or meaningless codes) is required.

Standard operating procedures (SOP), in which tasks are clearly defined and work steps are systematically described in detail, that is, who does what, when, and how, have been increasingly used on farms in recent years and contribute to improved data quantity and quality. Even if on-farm level SOPs are used for increasingly standardized data collection, there is often a lack of a common definition of diseases in veterinary practices with several employees.

In general, there is a lack of a uniform definition of diseases and production parameters that are applied across company borders, at national and international levels. Together with the variation in examination methods, this makes the comparison of, for example, disease incidences and key performance indicators between farms, regions, and countries, and, in science, of multicenter studies difficult or simply impossible. Even if increasing standardization in defining and reporting health data as well as production-related indices, for example, by published guidelines of the International Committee for Animal Recording (ICAR), is highly appreciated, there is still a need for further standardization as already requested for decades (Fetrow et al., 1990, 2006; Cook et al., 2006a; Pannwitz, 2015). One of the few examples of a cross-border standardization of health data recordings of dairy cows in a non-governmental database was developed by DSAHR Inc. (Association des médecins vétérinaires praticiens du Québec, St-Hyacinthe, Canada), which is available in Canada, France, Belgium, and Switzerland.

Considering animal health at the farm level, clinical disease is only the 'tip of the iceberg' underestimating the prevalence of disorders, which limit animal performance (LeBlanc, 2010). Furthermore, even subclinical forms of diseases, for example, subclinical ketosis, endometritis, hypocalcemia, and mastitis, can impair the health, well-being, and performance. Simply put, more accurate and cheaper measurement methods and diagnostic tools have led to this new understanding of diseases, where factors limiting performance at animal or herd level are considered as components of disease (LeBlanc et al., 2006). Currently, the routine and standardized examination of animals for the presence of subclinical diseases is recommended as an integral part of modern herd management (Cook et al., 2006a,b; LeBlanc et al., 2006). For this, several affordable diagnostic tools fostering herd management decisions have been developed.

### 3.2 National database

As a result of the BSE crisis at the end of the last century, Regulation No 1760/2000 was passed in Europe, which regulates the identification of cattle. For all bovines born on or after January 1, 1998, the regulation requires the lifelong application of two identical ear tags with a unique registration number and a passport enabling the animal to be traced back to the premises of birth. Furthermore, all animal movements must be documented in national databases. The unique identification and traceability of cattle enables the national and international monitoring of animal diseases and zoonosis so that, if necessary, measures can be taken as quickly as possible to prevent the outbreak or to implement control measures. Recent scenarios where databases were intensively used in Europe for tracking the spread of disease were the outbreak of Bluetongue starting in 2007 and Schmallenberg disease starting in 2011 (Wilson and Mellor, 2009; Koenraadt et al., 2014), which have led to the introduction of appropriate countermeasures.

Besides international monitoring of emerging and re-emerging infectious diseases, the unique identification of animals is a prerequisite for collecting reliable animal health data, at farm and national levels. In the Scandinavian countries, veterinary diagnoses have been collected and analyzed for a number of years (Nielsen et al., 2000; Forshell and Østerås, 2001; Philipsson and Lindhé, 2003; Negussie et al., 2010).

Under the leadership of the Federation of Austrian Cattle Breeders (ZAR, Vienna, Austria), a system for capturing veterinary diagnostic data in cattle was introduced throughout Austria in 2006. Various Austrian organizations involved in animal health, that is, the Ministry for Agriculture, Forestry, Environment and Water Management, the Ministry for Health, the University of Veterinary Medicine Vienna, the University of Natural Resources and Life Sciences Vienna, local animal health organizations, the Chamber of Agriculture, and the Chamber of Veterinarians, have contributed to further standardization of disease definition, data entry, and the development of the common database (Egger-Danner et al., 2012). Based on validated data sets of veterinary treatments, incidences in Simmental cows of 5.2% for metritis, 11.0% for mastitis, and 0.9% for ketosis were reported for 2010 (Egger-Danner et al., 2012). Additional information is presented in Figure 2.

An often-occurring problem is the underestimation of a disease at farm level. For instance, the above-mentioned incidence for subclinical and clinical ketosis in dairy cows based on veterinary treatments was 0.9%. This number, however, is not based on systematic monitoring of ketosis but on observations made by farmers or on occasional veterinary diagnoses. Particularly for subclinical diseases, this does not reflect the true incidence. This is illustrated by the fact that systematic ketosis monitoring on farms revealed incidences

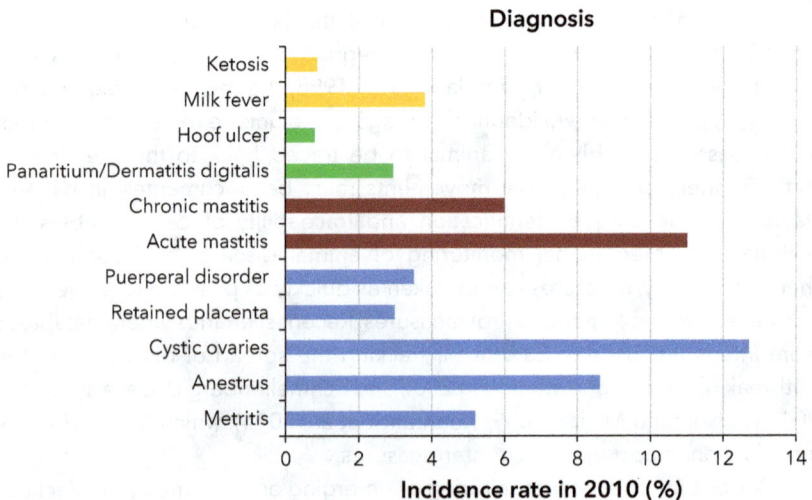

**Figure 2** Incidences for various cattle diseases in Austria 2010, based on recorded veterinary diagnoses and treatments (Egger-Danner et al., 2012).

between 11.2% and 36.6% in Europe (Suthar et al., 2013). With regard to subclinical ketosis, overseen cases do not necessarily result in obvious health problems for the animal but lead to a higher susceptibility to other diseases and to an impaired reproductive performance (Duffield et al., 2009; Chapinal et al., 2012).

# 4 On-farm diagnostic tests

In veterinary medicine and farm management, various tools, including diagnostic tests, are available for animal monitoring. In a broader sense, all (electronic) tools and devices that reduce the uncertainty about an animal's health and welfare status, as well as about its physiological state (e.g. pregnant vs. non-pregnant), can be considered as diagnostic tests. The variety of these tests is broad, ranging from a 'simple' combination of clinical examinations, which are indicative of a particular disease, to on-farm cow-side tests and the use of sophisticated laboratory methods. Examples of veterinary disciplines in which laboratory tests are used include clinical chemistry, cytology, hematology, histology, microbiology, serology, parasitology, toxicology, and fluid analyses. Describing the methodologies of these various tests in detail is beyond the scope of this chapter. About 30 years ago, the value of metabolic analyses, also with regard to their use in herd management, was recognized and triggered further activities in this field of research. In addition to technological developments, for example, cheaper devices for use on farms, this has led to an increasing interest of farmers, veterinarians, and consultants in using these tools, and it is expected that the 'rapid advancement of technologies that will enable real-time, automated measurement, will continue to revolutionize how we approach the monitoring of metabolic function and health in dairy cattle' (Overton et al., 2017). More recently and in addition to the previously described diagnostic approaches, farmers have increasingly been using sensor technologies for animal health monitoring, the potential of which should not be ignored by the veterinary profession. Several applications of sensor-based monitoring systems are merchandized by a rapidly increasing number of manufacturers and providers. In order to ensure that the technical tools and devices can be used successfully in veterinary practice as well as on dairy farms, however, they should be tested independently before commercialization with regard to their diagnostic reliability and economic benefits.

## 4.1 Validation and test performance of tools and devices

Several diagnostic tests used in veterinary medicine have been derived from human medicine. Even if the same parameter is measured, various species-specific and environmental factors can alter the results. It is therefore important that only tests that have been previously validated (Abuelo and Alves-Nores, 2016) for the species to be examined, preferably in multicenter studies, are used. Strictly speaking, validations are reliable only for a single population of animals, considering their housing conditions and production purpose (i.e. beef and/or milk production), for instance. This also applies to reference limits which are only valid for a specific laboratory with given equipment, for example, personnel, devices, and reagents (Lumsden et al., 1980; Iwersen et al., 2013).

An ideal test would be able to perfectly distinguish between the presence and absence of disease, without giving any false-positive or false-negative results. A good clinical test should 'provide correct diagnoses, provide data to aid in prognosis, provide an indication of subclinical disease, provide for monitoring the effects of treatment and provide data that may indicate possible reoccurrence of chronic disease' (Galen and Gambino, 1975, cited in Reneau 1986).

In this context, the terminology 'true positive' (TP), 'false positive' (FP), 'true negative' (TN), and 'false negative' (FN) refers to a comparison of the predicted condition, for example, the test result, with the true condition, for example, the animal status, as presented in Figure 3.

With regard to animal health, correct outcomes include the cases where an animal is either diseased and with a positive test result (TP), or non-diseased with a negative test result (TN). A Type I error occurs in the case of a positive test result (i.e. diseased) when the animal is actually healthy. A Type II error exists when the test indicates a healthy animal when it is actually diseased. A Type I error can cause financial losses due to unnecessary treatment, or in a worst-case culling, a Type II error can constitute an animal welfare issue, because

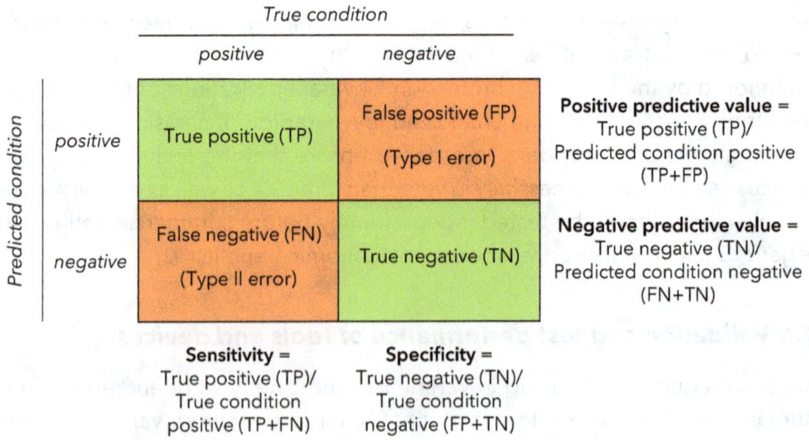

**Figure 3** Contingency table for the evaluation of a (diagnostic) test.

the diseased animals could have had the benefit of treatment or, in case of contagious diseases, the spread of disease. Hence, a possible treatment success leading to improved animal health and performance remains unrealized (Burfeind et al., 2010). Furthermore, an untreated animal bears a greater risk of becoming chronically diseased and/or spreading the disease in the herd.

For the quantification of the discriminate potential of a test system, sensitivities, specificities, as well as positive and negative predictive values are often reported in scientific validations. Detailed calculations of the above-mentioned test characteristics are presented in Figure 3. The diagnostic sensitivity (SE) is defined as the ability of a test to correctly classify an animal as diseased; diagnostic specificity (SP) is the ability of a test to correctly identify an animal as non-diseased. To determine SE and SP, the true condition of an animal must be known; that is, there must be a 'gold standard' (or a test combination) that always provides the correct information on the animal's condition. In practice, a gold standard does not always exist. Hence, sometimes methods that are believed as close enough to the true condition are considered as 'gold' (McKenna and Dohoo, 2006).

If the intention to test is to confirm a disease, one should use a test with a high SP, as there are few false positives. Otherwise, if the intention is to rule out disease, one should use a test with a high SE as there are few false negatives (Dohoo et al., 2009). The SE and SP of a test do not directly provide information on its performance when applied to animals with unknown disease status. The number of true and false test results is also dependent on the prevalence of the disease. The probability that an animal is or is not diseased, depending on whether the test is positive or not, is defined as the predictive value, which changes with different populations of tested animals (Dohoo et al., 2009). The positive predictive value (PPV) is the percentage of positive tested animals, which are actually suffering from the disease. The negative predictive value (NPV) is the percentage of animals with a negative test result, which do not have the disease.

Once again it should be mentioned that predictive values are dependent on the prevalence of the disease (see Tables 2 and 3). Hence, the direct comparison of the PPV and NPV of tests, captured in different populations and settings, is not valid. To get out of this dilemma, for comparison purposes, predictive values could be calculated for hypothetical levels of prevalence (Carrier et al., 2004; Iwersen et al., 2009).

### 4.1.1 Selecting a diagnostic test

In human medicine all clinical laboratory testing is regulated by federal legislation (Arnold et al., 2019). No equivalent regulation exists for veterinary medicine (Bertani et al., 2019). Therefore, independent analyses or approvals, as requested in human medicine from the manufacturers, are not required before introducing diagnostic tests into the veterinary market. The American Society for Veterinary Clinical Pathology (ASVCP) developed quality assurance guidelines for veterinary point-of-care tests (Flatland et al., 2013). Due to the lack of legal regulation, quality assurance (QA) and quality control of veterinary diagnostic tests should be in the own interest of the veterinary profession (Arnold et al., 2019).

The increase in diagnostic tests has led to various guidelines and recommendations toward testing procedures, and various publications on test performance were published. Evaluations of tests at farm conditions can reveal important features that were not revealed in optimal condition, for example, in a laboratory setting. For instance, test results may be influenced by the environment, the tested specimen, or the user's level of training (Iwersen et al., 2013; Pineda and Cardoso, 2015; Leal Yepes et al., 2018). Generic criteria which have to be considered besides SE and SP when selecting a diagnostic test are, the costs, user-friendliness, robustness, need of additional testing equipment, and time to perform the test, for example. However, these criteria need to be adapted to the specific needs of the farm and/or situation. Further information on monitoring and testing dairy herds can be found elsewhere (Oetzel, 2004; Cook et al., 2006a, b; Flatland et al., 2013).

## 4.2 Overview of available tests

In practice, it is often an advantage to have reliable test results available as quickly as possible, which are the basis for well-informed decisions, for example, for introducing preventive measures at herd level or for the treatment of individual animals. Next to feces, urine, blood, saliva, and rumen fluid, milk is an easily available specimen that has been used for decades in individual animal diagnostics.

An example of relatively inexpensive tests that have been used for decades in veterinary practice is the 'California Mastitis Test' (CMT) for (subclinical) mastitis detection (Schalm and Noorlander, 1957), sodium nitroprusside–based powders and tablets for diagnosing ketosis (Rothera, 1908), and a variety of liquid reagents, indicator papers, and test strips used for semi-quantitative measurement of single parameters, for example, ketones (Nielen et al., 1994; Geishauser et al., 2000; Osborne et al., 2002; Carrier et al., 2004), glucose (Rollin, 2006), calcium (Matsas et al., 1999; Fiore et al., 2020), pH, or a portfolio of metabolic parameters provided on one test strip (e.g. pH, specific gravity, leucocytes, nitrite, protein, glucose, ketones, bilirubin, blood). Many of the above-mentioned tests are for semi-quantitative analyses, which are based on increasing color intensity with increasing concentration of the substrate to be detected. Besides this, other relatively inexpensive instruments enabling quantitative measurement have become available for use in veterinary practices or on farms, for example, refractometer [reviewed by George (2001)] to analyze protein and urine solute concentration and to assess colostrum quality as well as the passive transfer of immunoglobulins in calves (McGuirk and Collins, 2004; Godden, 2008; Bartens et al., 2016).

Furthermore, on-farm tests for the rapid identification of infectious agents are on the market for many years, for example, for causative organisms of calf diarrhea (Luginbühl et al., 2005; Klein et al., 2009; Lichtmannsperger et al., 2019) and mastitis (Lago and Godden, 2018; Jones et al., 2019), or became recently of interest as a quick bacteriological test for a rough estimation of involved pathogens (Madoz et al., 2017).

## 5 Electronic devices and precision livestock farming technologies

### 5.1 Electronic devices

In recent years, various devices referred to as biosensors have been introduced in veterinary diagnostics and there is a growing number of hand-held devices (also referred to as point-of-care tests, on-farm tests, cow-side-tests) fostering animal health and herd management decisions. The first system that fits the definition of a biosensor, as provided later, was the introduction of a glucose oxidase sensor by Clark and Lyons (1962), a sensor which is, with modifications, still the most widely used (Grieshaber et al., 2008).

It is expected that in the veterinary diagnostics market point-of-care tests will increase at a compound annual growth rate of 18%, reaching US$ 6.71 billion by 2021 (Gattani et al., 2019).

Many of the hand-held devices used in veterinary medicine belong to the class of electrochemical biosensors, a subclass of biosensors, which convert biological information into a quantifiable electronic signal. In general, a biosensor consists of (a) a bio-receptor, which specifically binds to the analyte, (b) an interface architecture in which specific biological events take place and result in a signal that is picked up by, (c) the transducer element, converting the signal into an electronic signal which is amplified by a specific detector circuit enabling further processing, for example, by (d) computer software into a meaningful physical parameter, and (e) an interface with the operator where the quantified result is presented (Grieshaber et al., 2008). Elements used for bio-recognition of analytes typically consist of enzymes, nucleic acids, receptors, antibodies, and whole cells of which enzymes are most commonly used. Methods for the detection of changes in these biologic events often rely on amperometric (generating a measurable current), potentiometric (generating a measurable potential or charge accumulation), or conductometric (measurable alteration of conductive properties of a medium) procedures (Grieshaber et al., 2008).

Suitable on-farm devices can be used even under harsh conditions on farms, provide reliable results in a very short time and are relatively cheap. They usually require only small sample volumes and are easy to operate. The test media used ideally have a long shelf time, as, for example, for many dry chemistry tests. Furthermore, the provided (numerical) measures are easy to interpret, in particular, compared with color change–based semi-quantitative tests. In recent years, more and more devices evaluated in independent scientific studies often have significantly better test characteristics than the traditionally used test systems. These advantages lead to a steadily growing interest among farmers, veterinarians, and consultants in using these technologies.

## 5.2 Precision livestock farming technologies

Since the 1980s, sensors have been developed to capture parameters from individual farm animals. For dairy cows, the development started with individual cow identification and was followed by sensors that measured the electrical conductivity of milk for mastitis monitoring and pedometers to record the animal's activity for estrus detection. The introduction of automated milking systems in western Europe was accompanied by an increased focus on sensors to assist the farmer in herd and animal health management, leading to today's precision livestock technologies (Rutten et al., 2013). The core idea of precision livestock farming (PLF) is 'the use of information and communication technologies for improved control of fine-scale animal and physical resource variability to optimize economic, social, and environmental farm performance' (Eastwood et al., 2012). The continuous and automated monitoring of livestock aims to facilitate an early warning system and to optimize management procedures, leading to improved health and welfare of (individual) animals. While general herd management procedures in the past focused on larger groups or pens of animals, modern PLF technologies have the potential to pay more attention to individual animals within the herd. The techniques allow the creation of an individualized production management plan, a unique approach in livestock farming. The potential of PLF technologies to detect diseased animals earlier compared with classical diagnostic tools is important from an animal health and welfare perspective, as well as for the success of veterinary interventions. For managers of bigger farms, PLF technologies offer the opportunity to remove subjectivity from decision-making processes and reduce the need for skilled co-workers experienced in animal management (Borchers, 2015). Additionally, PLF technologies are able to monitor the feed efficacy, reducing waste of resources and minimizing the impact on the environment. For instance, a more targeted use of feedstuff and nutrients can lower phosphorus excretions. Besides maximizing production efficiency, PLF technologies reduce airborne pollution, improving the sustainability of livestock farming, and improving the occupational health of farm workers (Banhazi et al., 2010).

PLF technologies can continuously measure key parameters and provide quantitative information on the state of an animal, welfare, health, environmental load, or productivity. For animal monitoring various technologies are available, for instance, based on image, video, and audio analysis, or using innovative sensor systems attached to the animals (Berckmans, 2008; Berckmans and Guarino, 2008). In particular, video and audio technologies can monitor the animals without disturbing their natural behavior (Müller and Schrader, 2003). Sensors used in dairy cows can be divided into attached and nonattached systems. Attached sensors can either be placed inside the cow's body (in-cow), for example, as bolus or implant, or fixed on the cow's body (on-cow), for example, with collars or wristbands. Nonattached (off-cow) sensors were developed for measurements while cows pass by, through or over the device. Furthermore, nonattached sensors can be classified as in-line and on-line sensors. While in-line sensors were designed for continuous measurement of a fluid (e.g. milk) flowing through a pipe (e.g. in the milk line), on-line sensors take a sample for analyses (Rutten et al., 2013). From a practical and scientific point of view, the opportunity to collect big data using PLF technologies has the potential to provide knowledge on biology, physiology, and animal behavior at a far greater intensity than in the past. The automated milking and herd (health) monitoring systems increasingly used on farms are sources of big data that can be used by various stakeholders (e.g. farmers, veterinarians, consultants, companies). Inter alia, methods of artificial intelligence are used here to identify specific patterns in the data, which are associated with certain animal behaviors or diseases. At this point it should be pointed out that the knowledge newly generated from the multitude of farm data should be made available to the farmers for further business development.

Various PLF systems are on the market. The sheer size of the livestock market makes it possible to produce customized technologies at low unit cost (Berckmans, 2014). Most of the systems, however, have not yet been shown to give a clear benefit under different operating conditions. Research has frequently been conducted at the laboratory level or on research and/or single farms. There has been an emphasis on developing sophisticated sensor technologies, that is, the hardware, but the setup of a resilient background for the systems is often neglected, and there is a lack of customer service and training, easy to understand manuals, a reliable data storage and backup system, and user-friendly software (Rutten et al., 2013). Due to the lack of PLF systems installed on the farm level, it is hardly possible to evaluate the value or benefit of PLF technology. To date, the development of PLF technologies has largely been driven by European companies and the PLF systems were predominately used by European farmers (Banhazi et al., 2012), but currently there is a growing interest worldwide in the use and improvement of these technologies. The desire to increase milk production per cow while decreasing production costs is one of many reasons that farmers adopt technologies (De Koning, 2010; Steeneveld et al., 2012). PLF is one of the most powerful developments in modern farming, with the potential to revolutionize the livestock farming industry and improve the economic stability of rural areas.

# 6 Case study: detecting subclinical ketosis in dairy cows

Subclinical ketosis in early lactation is a widespread metabolic disease in dairy cows, which is often underestimated by farmers and/or veterinarians. Elevated concentrations of non-esterified fatty acids (NEFAs) and ketone bodies in blood in the absence of clinical symptoms are characteristic of the disease. The concentrations of NEFA and the ketone ß-hydroxybutyrate (BHB) are directly proportional to body fat mobilization, which is needed to support milk production in early lactation. A concentration of these metabolites beyond physiological limits is associated with an increased risk for sequalae (e.g., clinical ketosis, displaced abomasum, mastitis, metritis), decreased milk yield, and impaired reproductive performance, affecting the economics of a dairy farm (Sturm et al., 2020). Because of the subclinical nature of the disease, in particular, the absence of clinical symptoms, a routine testing of dairy cows in early lactation is recommended as part of a modern herd health management procedure.

Due to the widespread nature of this often underestimated disease and its influence of the economic success of the farm, the following chapter will further discuss the disease and its on-farm diagnosis, including PLF technologies.

## 6.1 Case definition, prevalence, and impact on animal health and performance

As presented in Figure 4, in early lactation, the high energy requirements of milk production accompanied by a delayed increase in dry matter intake often lead to a negative energy balance (NEB) in dairy cows (Ingvartsen and Andersen, 2000; Grummer et al., 2004). According to Villa-Goday et al. (1988), approximately 80% of cows go through an at least temporary phase of NEB. In their review of 49 studies on energy balance published in 20 peer-reviewed scientific journals, Grummer and Rastani (2003) reported an average duration of NEB of 45 days (ranging from 7 to 105 days, standard deviation 21 days), with 90% of animals being in a positive energy balance by 63 days post-partum.

Successful adaptation to NEB is of central importance for ensuring animal health, high levels of reproductive performance, and milk yield (Drackley,

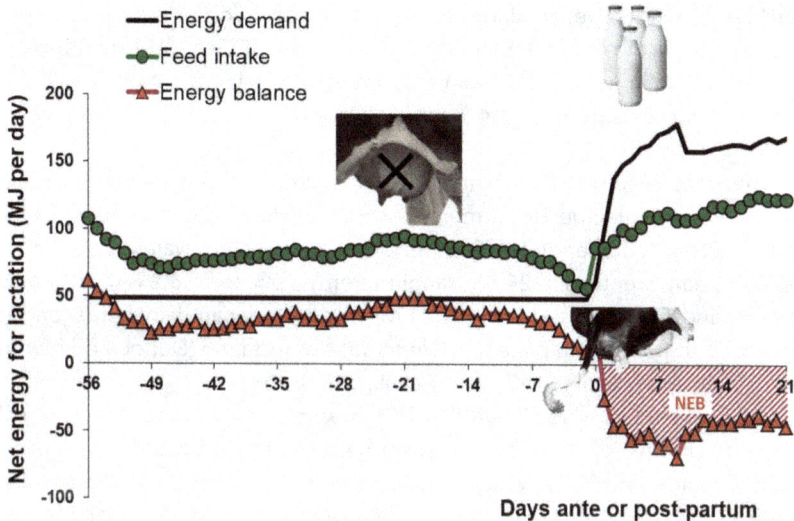

**Figure 4** Example of energy (MJ $NE_L$ per day) requirements, consumption, and the resulting energy balance during an experiment by Grummer (2008). Modified graph.

1999; Herdt, 2000; Ingvartsen, 2006). A maladaptation to this phase leads to excessive mobilization of body fat with increasing concentrations of NEFAs. If the metabolic capacity of the liver's hepatocytes is exceeded, ketogenesis occurs, leading to increased concentrations of ketones, that is, acetate (Ac), acetoacetate (AcAc), and ß-hydroxybutyrate in blood, urine, milk, and cerebrospinal fluid and other body fluids.

These increased concentrations of ketones lead to the disease being known as 'ketosis,' also referred to as acetonemia or hyperketonemia. This metabolic disease can occur not only as clinical ketosis but also as subclinical ketosis (SCK) (Duffield, 2000), that is, in the absence of any clinical symptoms, such as central nervous disorders, including salivation, dullness, and aggression (Andersson, 1988).

Acetoacetate decomposes rapidly to acetone and carbon dioxide and is therefore considered an unstable compound of low diagnostic value (Bergman, 1971; Bruss, 1989). In contrast to AcAc, BHB is a relatively stable compound, both in vivo and in vitro (Tyopponen and Kauppinen, 1980; Custer et al., 1983). Hence, the analysis of BHB in serum or plasma is recommended as the gold standard test for diagnosing ketosis not only at the animal level but also at the herd level (Duffield, 2000). Recommended thresholds to define SCK in cows are BHB concentrations in blood of 1.2 mmol/L and 1.4 mmol/L (Geishauser et al., 1998; Duffield et al., 2009), which have been used as a reference in recent research. Suthar et al. (2013) suggest the use of a uniform threshold to define SCK of 1.2 mmol/L BHB in blood. From the herd level perspective, an alarm level (i.e. the proportion of sampled animals above a certain threshold) for animals suffering from SCK of 15% in multiparous cows and 20% in heifers has been recommended, using a cut-off of the BHB concentration in serum of 1.4 mmol/L (Cook et al., 2006a; Ospina et al., 2010).

The occurrence of SCK in dairy cows in the periparturient period represents an important challenge for farmers. Based on BHB concentrations in blood of ≥1.2 mmol/L, Suthar et al. (2013) reported an overall prevalence of SCK for ten European countries of 21.8%, ranging from 11.2% to 36.6% within 2 weeks after calving. The reported prevalence for North American dairy herds ranged from 8.9% to 43.2% within the first two months of lactation (Dohoo and Martin, 1984; Geishauser et al., 1998; McArt et al., 2012).

Several studies have shown that SCK is associated with an increased risk of the occurrence of secondary diseases (LeBlanc et al., 2005; Duffield et al., 2009; Suthar et al., 2013). Additionally, a decrease in milk yield (Dohoo and Martin, 1984; Duffield, 1997) and impaired reproductive performance (Walsh et al., 2007; Chapinal et al., 2012) are associated with the occurrence of SCK. Raboisson et al. (2014) conducted a meta-analysis on 23 publications on the occurrence of sequalae related to SCK. In the presence of SCK (BHB >1.4 mmol/L in the blood), the authors determined relative risks for displaced abomasum, lameness, metritis, clinical mastitis, and retained fetal membranes of 3.33, 2.01, 1.75, 1.61, and 1.52, respectively.

Furthermore, an increased BHB concentration in pre-calving cows has been associated with a detrimental impact on milk yield in the next lactation and on animal health (Chapinal et al., 2011, 2012). Animals showing a BHB concentration in the serum of ≥0.7 mmol/L within the last week of gestation were at a greater risk of early culling (Roberts et al., 2012).

## 6.2 Advances in ketosis monitoring

Considering the impact of hyperketonemia on animal health, welfare, and performance, monitoring of dairy herds for SCK is an appropriate measure for disease prevention and improvement of herd management efficiency in dairy farming (Cook et al., 2006b). Helpful guidelines, defining, for example, thresholds, alarm levels, and the animals of interest for metabolic testing, as presented in Table 1, have been published (Stokol and Nydam, 2005; Cook et al., 2006a,b).

However, the quantitative determination of BHB as the gold standard depends on special laboratory analyzers and requires blood sampling, centrifugation, freezing of plasma or serum samples, and shipping of frozen specimens to the laboratory. In order to minimize these inconveniences, in particular, to provide results immediately after sampling on farm and to reduce

**Table 1** Diagnostic tests, cut-off points, and alarm levels used in animals at risk to identify subclinical diseases

| Test | Cut-off point | Alarm level proportion | At-risk group | Associated disease |
|---|---|---|---|---|
| ß-Hydroxybutyrate (BHB) | ≥1.4 mmol/L | >10% | Lactating cows 50 to 50 DIM | Ketosis, displaced abomasum |
| Non-esterified fatty acid (NEFA) | ≥0.4 mmol/L | >10% | Dry cows 2 to 14 days before expected calving | Ketosis, displaced abomasum, fatty liver disease |
| Ruminal pH | ≤5.5 | >25% | Lactating cows 50 to 50 DIM in herds fed concentrate separately, 50 to 150 DIM in herds fed TMR | Subacute ruminal acidosis |
| Blood calcium | ≤2.0 mmol/L | >30% | Lactating multiparous cows 12 to 24 hours after calving | Clinical milk fever |

DIM: Days in milk, TMR: total mixed ration. Source: Cook et al. (2006a).

laboratory costs, various cow-side diagnostic tests, based on the reaction of AcAc and Ac with sodium nitroprusside leading to a ketone-dependent color change (Rothera, 1908; Adler et al., 1957), have been developed. Higher levels of ketones lead to a more intensive coloration; hence, these tests can be regarded as semi-quantitative.

### 6.2.1 Traditional semi-quantitative tests

To this day, these semi-quantitative tests are still used in the daily practice of farmers and veterinarians as tablets or dipsticks to detect AcAc, and to a lesser degree Ac in urine (e.g. Ketostix®, Bayer Corp.; AimTab™ Ketone Tablets, Germaine) or BHB in milk (PortaBHB®, PortaCheck Inc.; Keto-Test™, Elanco). An overview of scientifically evaluated semi-quantitative tests for the determination of ketones in milk and urine, respectively, are presented in Tables 2 and 3. An example of their application is shown in Figure 5.

Differences in SE, SP, and predictive values may be caused by variation in sampling or in the error-prone subjective interpretation of the color changes of the test strips (Carrier et al., 2004). Moreover, it has been demonstrated that high levels of somatic cell count (SCC) in milk (Jeppesen et al., 2006) and feeding of mal-fermented silages (Cook et al., 2006b) can distort the results of

**Table 2** Performance of different commercially available cow-side tests for use in milk for detecting subclinical ketosis at different cut-off levels

| Reference | Test and cut-offs (mmol/L) | Serum (mmol/L) | n[1] | SE[2] (%) | SP[3] (%) | PV$_{15}$[4] (%) | | PV$_{30}$[5] (%) | |
|---|---|---|---|---|---|---|---|---|---|
| | | | | | | pos | neg | pos | neg |
| Geishauser (1998) | Ketolac[6] | ≥1.2 | 529 | | | | | | |
| | ≥0.05 | | | 92 | 55 | 27 | 97 | 47 | 94 |
| | ≥0.1 | | | 72 | 89 | 54 | 95 | 74 | 88 |
| | ≥0.2 | | | 45 | 97 | 73 | 91 | 87 | 80 |
| | ≥0.5 | | | 17 | 100 | 100 | 87 | 100 | 74 |
| | ≥1.0 | | | 3 | 100 | 100 | 85 | 100 | 71 |
| Geishauser et al. (2000) | Ketolac | ≥1.4 | 469 | | | | | | |
| | ≥0.05 | | | 91 | 56 | 27 | 97 | 47 | 94 |
| | ≥0.1 | | | 80 | 76 | 37 | 96 | 59 | 90 |
| | ≥0.2 | | | 59 | 90 | 51 | 93 | 72 | 84 |
| | Pink[7] | | | | | | | | |
| | ≥0.1 | | | 76 | 93 | 66 | 96 | 82 | 90 |
| | ≥0.2 | | | 38 | 98 | 77 | 90 | 89 | 79 |
| | Uriscan[8] | | | | | | | | |
| | ≥500 | | | 13 | 100 | 100 | 87 | 100 | 73 |
| | Rapignost[9] | | | | | | | | |
| | ≥500 | | | 3 | 100 | 100 | 85 | 100 | 71 |
| Osborne et al. (2002) | KetoTest[10] | ≥1.4 | 121 | | | | | | |
| | ≥0.1, 1 wk pp | | | 95 | 71 | 37 | 99 | 58 | 97 |

| Study | Test | Threshold | n | | | | | | |
|---|---|---|---|---|---|---|---|---|---|
| Carrier et al. (2004) | ≥0.1, 2 wk pp | | 127 | 95 | 67 | 34 | 99 | 55 | 97 |
| | KetoTest | ≥1.4 | 850 | | | | | | |
| | ≥0.05 | | | 88 | 90 | 61 | 98 | 79 | 95 |
| | ≥0.1 | | | 73 | 96 | 76 | 95 | 89 | 89 |
| | ≥0.2 | | | 27 | 99 | 83 | 88 | 92 | 76 |
| | ≥0.5 | | | 3 | 100 | 100 | 85 | 100 | 71 |
| | ≥1.0 | | | 2 | 100 | 100 | 85 | 100 | 70 |
| | KetoCheck[11] | ≥1.4 | 845 | | | | | | |
| | ≥Trace | | | 41 | 99 | 88 | 90 | 95 | 80 |
| | ≥Moderate | | | 10 | 100 | 100 | 86 | 100 | 72 |
| | ≥Large | | | 2 | 100 | 100 | 85 | 100 | 70 |
| Iwersen et al. (2009) | Ketolac[12] | ≥1.4 | 194 | | | | | | |
| | ≥0.1 | | | 90 | 94 | 73 | 98 | 87 | 96 |
| | ≥0.2 | | | 30 | 98 | 73 | 89 | 87 | 77 |

[1] Number of observations paired with a serum BHB measurement for each cow-side test.
[2] SE = Sensitivity: proportion of diseased cows that test positive.
[3] SP = Specificity: proportion of non-diseased cows that test negative.
[4] Positive and negative predicted values based on a theoretical prevalence of subclinical ketosis of 15%.
[5] Positive and negative predicted values based on a theoretical prevalence of subclinical ketosis of 30%.
[6] Hoechst, Unterschleißheim, Germany.
[7] Profs-Products, Wittibreut, Germany.
[8] Heiland, Hamburg, Germany.
[9] Behring, Marburg, Germany.
[10] Sanwa Kagaku, Nagoya, Japan.
[11] Great States Animal Health, St. Joseph, United States of America.
[12] Biolab, München, Germany.

**Table 3** Performance of different commercially available cow-side tests for use in urine for detecting subclinical ketosis at different cut-off levels

| Reference | Test and cut-offs (mmol/L) | Serum (mmol/L) | n[1] | SE[2] (%) | SP[3] (%) | PV[4] 15 (%) | | PV[5] 30 (%) | |
|---|---|---|---|---|---|---|---|---|---|
| | | | | | | pos | neg | pos | neg |
| Nielen et al. (1994) | Acetest[6] | ≥1.2 | 124 | 91 | 61 | 29 | 97 | 50 | 94 |
| | | ≥1.4 | 124 | 100 | 59 | 30 | 100 | 51 | 100 |
| Osborne et al. (2002) | KetoTest[7] | ≥1.4 | | | | | | | |
| | ≥0.1, 1 wk pp | | 71 | 93 | 54 | 26 | 98 | 46 | 95 |
| | ≥0.1, 2 wk pp | | 88 | 100 | 65 | 34 | 100 | 55 | 100 |
| Carrier et al. (2004) | Ketostix[8] | ≥1.4 | 710 | | | | | | |
| | ≥0.5 | | | 90 | 86 | 53 | 98 | 73 | 95 |
| | ≥1.5 | | | 78 | 96 | 77 | 96 | 89 | 91 |
| | ≥4.0 | | | 49 | 99 | 90 | 92 | 95 | 87 |
| | ≥8.0 | | | 12 | 100 | 100 | 87 | 100 | 73 |
| | ≥16.0 | | | 4 | 100 | 100 | 86 | 100 | 71 |
| Iwersen et al. (2009) | Ketostix[9] | ≥1.4 | 186 | | | | | | |
| | ≥0.5 | | | 78 | 92 | 63 | 96 | 81 | 91 |
| | ≥1.5 | | | 67 | 97 | 80 | 94 | 91 | 87 |
| | ≥4.0 | | | 67 | 100 | 100 | 94 | 100 | 88 |

| | | | | | | | |
|---|---|---|---|---|---|---|---|
| ≥8.0 | 44 | 100 | 100 | 100 | 91 | 100 | 87 |
| ≥16.0 | 22 | 100 | 100 | 100 | 88 | 100 | 75 |

[1] Number of observations paired with a serum BHB measurement for each cow-side test.
[2] SE = Sensitivity: proportion of diseased cows that test positive.
[3] SP = Specificity: proportion of non-diseased cows that test negative.
[4] Positive and negative predicted values based on a theoretical prevalence of subclinical ketosis of 15%.
[5] Positive and negative predicted values based on a theoretical prevalence of subclinical ketosis of 30%.
[6] Ames Devision, Glamorgan, United Kingdom.
[7] Sanwa Kagaku, Nagoya, Japan.
[8] Bayer Corporation, Elkhart, United States.
[9] Bayer, Leverkusen, Germany.

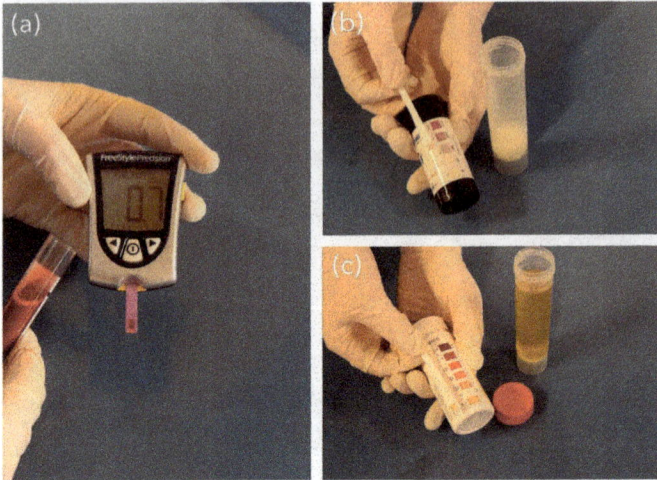

**Figure 5** Example for ketosis diagnostics by the use of (a) an electronic hand-held device in blood and semi-quantitative dipsticks for use in (b) milk and (c) urine.

milk ketone tests. Prolonged reaction times of the reagent with urine might also lead to FP results (Oetzel, 2004).

Even if the test characteristics vary between tests and studies, there is a general agreement that semi-quantitative tests are useful for ketosis monitoring, in particular, on herd level. Urine sampling, however, is often challenging, and the suboptimal specificities for urine testing and the inadequate sensitivities of milk testing were reported as drawbacks of the test systems (Oetzel, 2004).

### 6.2.2 Electronic hand-held devices and factors influencing the results

Electronic hand-held devices for measuring glucose (glucometer) and ketones (ketometer) are widely used in human medicine for diabetes monitoring (Guerci et al., 2005; Grieshaber et al., 2008). Ketometer systems consist of a hand-held meter and electrochemical test strips for BHB measurement (Figure 5). After inserting the test strip into the strip port of the device, the front edge of the application zone is brought into contact with, for example, a blood drop or serum sample. The specimen is drawn into the test strip, and for many devices, the chemical reaction starts as follows: the BHB in the specimen is oxidized to AcAc in the presence of the enzyme ß-hydroxybutyrate dehydrogenase, with the concomitant reduction of $NAD^+$ to NADH. The NADH is reoxidized to $NAD^+$ by a redox mediator. Electrons are released by this reaction leading to a small current, which is directly proportional to the BHB concentration in the specimen (Iwersen et al., 2009). For most devices, the BHB concentration is displayed as mmol/L within 10 seconds after specimen application on the test strip. For many devices, the minimum operating temperature is +4°C. This can lead to challenges when using the device during winter months. In this case, the device should be stored in a warm place (e.g. in a pocket near the body) until use (Iwersen et al., 2013).

Introducing ketometer for BHB measurements on farms has led to significant improvement in ketosis monitoring in dairy cows (Overton et al., 2017). In particular, presenting the results as a numerical value on a display and exact adherence to reaction times have improved the test interpretation and, hence, reduced the risk of misinterpretation, as reported for color-based semi-quantitative tests. Furthermore, only small amounts (0.8–1.5 µL) of whole blood, serum, or plasma are needed for measurements, and, in the case of using whole blood, further processing of samples (e.g. centrifugation, pipetting, storage) is not needed. The individually foil-packaged test strips meet the often unfavorable hygienic conditions on farms and reduce the potential negative impact on the quality of the test strips. If newer devices are used, calibration procedures prior to use are no longer necessary (Voyvoda and Erdogan, 2010; Iwersen et al., 2013).

Jeppesen et al. (2006) were among the first to study the use of an electronic hand-held device for BHB measurement (MediSense Precision, Abbott, Abingdon, UK) in dairy cows. The authors reported a high correlation of $r^2 = 0.99$, with spectrophotometrically determined BHB concentrations, which served as the gold standard in their study, and considered the test eligible for the monitoring of SCK in dairy cows. In recent decades, several hand-held devices have been tested under different farm conditions. For instance, in a survey of an electronic hand-held device tested by 35 veterinary practitioners in 77 farms, an SE of 96% and SP of 97% (based on a cut-off of 1.4 mmol/L BHB in serum) were reported (Iwersen et al., 2009). In the final evaluation by veterinarians (n = 30), 93% rated the diagnostic value of the device as 'good' or 'very good.' Of the respondents, 63% reported that they 'definitely' wanted to continue using the device and 23% would 'probably' continue using it. Today, for routine ketosis monitoring at animal and herd level, the use of hand-held electronic devices has become standard (Overton et al., 2017). An overview of evaluated hand-held devices for determining BHB in whole blood is presented in Tables 4 and 5. Reports on using other specimens for BHB testing using hand-held electronic devices are available elsewhere (Iwersen et al., 2009, 2013; Pineda and Cardoso, 2015).

In contrast to almost constant laboratory conditions, on-farm tests are used under varying environments, such as different temperatures in summer and winter. Therefore, various studies were carried out on the practical applications of hand-held electronic devices, for example, on the type of specimen, timing, and sampling procedures. Iwersen et al. (2013) determined

**Table 4** Performance of different commercially available hand-held devices for use with whole blood for detecting subclinical ketosis at a serum or plasma ß-hydroxybutyrate level of ≥1.2 mmol/L

| Reference | Device | Device cut-off[1] (mmol/L) | n | SE[2] (%) | SP[3] (%) |
|---|---|---|---|---|---|
| Iwersen et al. (2009b) | Precision Xtra[4] | 1.2 | 926 | 88 | 96 |
| Voyvoda and Erdogan (2010) | Optimum Xceed[4] | 1.2 | 78 | 85 | 94 |
| Panousis et al. (2011) | Precision Xceed[4] | 1.2 | 163 | 91 | 96 |
| Iwersen et al. (2013) | FreeStyle Precision[4] | 1.2 | 425 | 98 | 90 |
|  | GlucoMen LX Plus[5] | 1.1 | 415 | 80 | 87 |
| Mahrt et al. (2014a) | Precision Xtra | 1.2 | 92 | 90 | 88 |
| Mahrt et al. (2014b) | NovaVet[6] | 1.2 | 155 | 97 | 82 |
| Kanz et al. (2015) | FreeStyle Precision | 1.2 | 240 | 100 | 93 |
|  | GlucoMen LX Plus | 1.0 | 240 | 94 | 85 |
|  | NovaVet | 1.0 | 240 | 100 | 83 |
| Bach et al. (2016) | Precision Xtra | 1.2 | 89 | 100 | 74 |
|  | TaiDoc[7] | 1.2 | 89 | 100 | 74 |
|  | NovaMax[6] | 1.2 | 89 | 75 | 100 |
|  | NovaVet | 1.2 | 89 | 95 | 92 |
| Süss et al. (2016) | FreeStyle Precision Neo[4] | 1.1 | 240 | 100 | 95 |
| Macmillan et al. (2017) | FreeStyle Precision Neo | 1.2 | 441 | 98 | 95 |
| Zakian et al. (2017) | Precision Xtra[5] | 1.2 | 181 | 94 | 94 |
| Leal Yepes et al. (2018) | Precision Xtra | 1.2 | 100 | 100 | 87 |
|  | TaiDoc | 1.2 | 100 | 100 | 76 |
| Khol et al. (2019) | WellionVet BELUA[8] | 1.2 | 243 | 89 | 99 |

[1] (Optimized) device threshold used for diagnosing subclinical ketosis.
[2] SE = Sensitivity: proportion of diseased cows that test positive.
[3] SP = Specificity: proportion of non-diseased cows that test negative.
[4] Abbott Diabetes Care Ltd., Witney, United Kingdom.
[5] A. Menarini, Vienna, Austria.
[6] Nova Biomedical, Waltham, United States.
[7] Pharmadoc, Lüdersdorf, Germany.
[8] MED TRUST Handels GmbH, Marz, Austria.

**Table 5** Performance of different commercially available hand-held devices for use with whole blood for detecting subclinical ketosis at a serum or plasma ß-hydroxybutyrate level of ≥1.4 mmol/L

| Reference | Device | Device cut-off[1] (mmol/L) | n | SE[2] (%) | SP[3] (%) |
|---|---|---|---|---|---|
| Iwersen et al. (2009) | Precision Xtra[4] (study 1) | 1.4 | 196 | 100 | 100 |
|  | Precision Xtra (study 2) | 1.4 | 926 | 96 | 97 |
| Voyvoda and Erdogan (2010) | Optimum Xceed[4] | 1.4 | 78 | 90 | 98 |
| Panousis et al. (2011) | Precision Xceed[4] | 1.4 | 163 | 100 | 100 |
| Iwersen et al. (2013) | FreeStyle Precision[4] | 1.4 | 425 | 100 | 97 |
|  | GlucoMen LX Plus[5] | 1.3 | 415 | 86 | 96 |
| Mahrt et al. (2014a) | Precision Xtra | 1.4 | 92 | 89 | 90 |
| Mahrt et al. (2014b) | NovaVet[6] | 1.3 | 155 | 96 | 85 |
|  | NovaVet | 1.4 | 155 | 91 | 89 |
| Kanz et al. (2015) | FreeStyle Precision | 1.4 | 240 | 100 | 96 |
|  | GlucoMen LX Plus | 1.3 | 240 | 85 | 97 |
|  | NovaVet | 1.3 | 240 | 90 | 98 |
| Zakian et al. (2017) | Precision Xtra | 1.4 | 181 | 93 | 95 |

[1] (Optimized) device threshold used for diagnosing subclinical ketosis.
[2] SE = Sensitivity: proportion of diseased cows that test positive.
[3] SP = Specificity: proportion of non-diseased cows that test negative.
[4] Abbott Diabetes Care Ltd., Witney, United Kingdom.
[5] A. Menarini, Vienna, Austria.
[6] Nova Biomedical, Waltham, United States.

similar BHB concentrations when either whole blood or EDTA-added whole blood was tested. Furthermore, similar results have also been reported in BHB measurements using serum and plasma. However, because BHB concentrations in serum and plasma are lower compared with whole blood, specimen and device-specific thresholds should be used for ketosis monitoring (Iwersen et al., 2013; Pineda and Cardoso, 2015). The influence of different temperatures during device and test strip storage or sample measurement, respectively, in a temperature range between +5°C and +32°C have been investigated (Iwersen et al., 2013). The storage conditions of the devices and test strips did not affect the test results, but analyzing a blood sample at different temperatures caused significant differences. Hence, to obtain reliable results, the sample temperature should be close to body temperature (i.e. immediately after sampling) while testing; otherwise, the temperature of the sample has to be considered. The effect of the sampling time and sampling location (tail vessel, jugular vein, and mammary vein) on BHB concentrations was tested by Mahrt et al. (2014). The authors reported that the sampling time in continuously fed dairy cows did not affect BHB concentrations. Mean BHB concentrations in the mammary vein were 0.3 mmol/L and 0.4 mmol/L lower compared with concentrations in the jugular vein and tail vessel, respectively. Hence, the authors concluded that blood samples could be taken at any time of the day in continuously fed cows, preferably from the tail vessels or jugular vein; the mammary vein should not be used. Other studies investigated whether capillary blood, obtained minimally invasively by the use of a lancet on the skin of the exterior vulva (Kanz et al., 2015) or the edge of an ear (Süss et al., 2016), was suitable for ketosis testing using hand-held electronic devices. The authors concluded that capillary blood is eligible for classifying cows suffering from SCK, but thresholds should be adjusted. Applying these minimally invasive procedures, farmers have the possibility of testing their cows using electronic devices in countries where national legislation prohibits conventional blood sampling by laypersons.

### 6.2.3 Milk composition

Both milk fat and protein composition are significantly influenced by hyperketonemia (Miettinen and Setala, 1993; Duffield et al., 1997), and, thus, the fat-to-protein ratio (F/P), as well as the protein-to-fat ratio, has been used as indicators for SCK (Grieve et al., 1986; Duffield et al., 1997; Heuer et al., 1999; Jenkins et al., 2015). Monthly reports from dairy herd improvement associations (DHIA) provide the farmer with individual animal information on milk yield, components (e.g. fat, protein, lactose, urea), and SCC, which provide a rough overview of the metabolic status at the herd level (Nelson and Redlus, 1989; Hamann and Krömker, 1997; Cook et al., 2006a). Because of inadequate SE and SP, fat and protein percentages alone or a combination thereof and their ratios were considered as not being useful in detecting SCK in individual animals (Duffield et al., 1997; Jenkins et al., 2015). Furthermore, because of the long sampling intervals of approximately 4 weeks and the limited factors analyzed in composite milk samples, DHIA records are considered as ineffective as a basis for modern health management (Hamann and Krömker, 1997).

Several validations of in-line methods for analyses of ketones in routine milk analyses have been performed (Nielsen et al., 2005). One of the modern methods within the quantitative analysis is Fourier-transform infrared (FTIR) spectrometry, which is currently used as a standard method for multi-component milk testing (Hansen, 1999; Pralle and White, 2020). De Roos et al. (2007) reported a correlation of approx. 80% between FTIR predicted Ac and BHB concentrations compared with their respective chemical tests. The authors considered the prediction of Ac and BHB by FTIR as 'valuable for screening cows for ketosis' with a reported SE for BHB of 69% and Ac of 70%, and SP of 95% for both. Similar results, presented in Table 6, were reported by Van Knegsel et al. (2010), Van der Drift et al. (2012), and Denis-Robichaud et al. (2014).

In a recent study, King et al. (2019) tested the accuracy of in-line F/P to detect SCK in the first 3 weeks of lactation on 484 cows kept in nine herds with automated milking systems. Blood samples (n = 1427) were taken at weekly intervals, and BHB concentrations of ≥1.2 mmol/L and ≥1.4 mmol/L, respectively, were used to define SCK. Using various F/P ratios to detect SCK, the authors reported poor SE (58% to 92%) and SP (65% to 69%), with high rates of false positives and negatives ranging from 31% to 39%. Similar proportions of FP- and FN-classified animals are reported in the studies presented in Table 6. Because of low PPV, Van Knegsel et al. (2010) had 'concerns about the practical applicability of FTIR predictions of acetone, BHB, and fat-to-protein ratio in milk to detect hyperketonemic cows.' Other authors considered the method an 'accurate diagnostic tool' (Denis-Robichaud et al., 2014) or as suitable for the pre-selection of animals to be subjected to further testing for SCK (Hansen, 1999; Van der Drift et al., 2012; Petersson et al., 2017).

In-line systems for use on farms, for example, in automated milking systems or parlors, are available for supporting herd health management. Two systems that have been commercially available for some time, the AfiLab (Afimilk, Kibbutz Afikim, Israel) and Herd Navigator (DeLaval, Tumba, Sweden), are presented here as examples.

The AfiLab is an optical in-line milk analyzer, which measures milk fat, protein, lactose, blood admixture in milk in real time and the coagulation potential in every milking of an individual cow for detecting, for example, ketosis, mastitis, and acidosis. Analyses are based on multivariate analysis of near-infrared spectra during milking, using algorithms to predict the total amount of milk components. In this system, an F/P ratio of 1.4 is used for pre-selecting animals that need further testing for ketosis, for example, by measuring blood or urine (Afikim, 2019).

The Herd Navigator automatically takes representative milk samples from specific cows during milking intended to be used, for example, for reproduction, ketosis, and mastitis management. Prior to milking, an algorithm selects individual cows and specific parameters to be tested. The fully automated analytic device, which is installed in a separate place, relies on dry stick analyses, using colorimetric reactions for the determination of lactate dehydrogenase, urea, and BHB and an immuno-assay for measuring progesterone (Mazeris, 2010). For both systems, analytical outputs can be presented as tables and graphs in herd management software. Thresholds, for example, used for generating alarms, can be adjusted by the user.

Although in-line systems have been used on farms for some years now, independent publications demonstrating the benefit of their use in animal

**Table 6** Performance of Fourier-transform infrared (FTIR) spectroscopy for predicting hyperketonemia from testing milk samples at different cut-off levels

| Reference | Parameter[1] | Threshold[2] | n[3] | Prev[4] (%) | Test characteristics (%) | | | |
|---|---|---|---|---|---|---|---|---|
| | | | | | SE[5] | SP[6] | PPV[7] | NPV[8] |
| De Roos et al.[9] (2007) | BHB (µmol/L) | ≥100 | 1080 | 17.2 | 69 | 95 | 75 | 93 |
| | Ac (µmol/L) | ≥150 | | | 70 | 95 | 73 | 94 |
| Van Knegsel et al.[10] (2010) | BHB (µmol/L) | ≥23 | 69 | 7.1 | 80 | 71 | 18 | 98 |
| | Ac (µmol/L) | ≥70 | | | 80 | 70 | 17 | 98 |
| | F/P | ≥1.5 | | | 66 | 71 | 15 | 96 |
| Van der Drift et al.[10] (2012) | BHB (µmol/L) | ≥76.3 | 1678 | 11.2 | 83 | 76 | 30 | 97 |
| | Ac (µmol/L) | ≥131.5 | | | 71 | 89 | 45 | 96 |
| Denis-Robichaud et al.[11] (2014) | BHB (µmol/L) | ≥200 | 163 | 21.0 | 84 | 96 | 84 | 96 |
| | Ac (µmol/L) | ≥80 | | | 87 | 95 | 83 | 96 |
| | F/P | ≥1.3 | | | 69 | 66 | 33 | 90 |

[1] BHB = β-hydroybutyrate, Ac = Acetone, F/P = milk fat-to-protein ratio.
[2] (Optimized) device threshold used for diagnosing subclinical ketosis.
[3] Number of samples.
[4] Prev = Prevalence.
[5] SE = Sensitivity: proportion of diseased cows that test positive.
[6] SP = Specificity: proportion of non-diseased cows that test negative.
[7] PPV = Positive predictive value based on reported study prevalence.
[8] NPV = Negative predictive value based on reported study prevalence.
[9] Definition of subclinical ketosis based on BHB and Ac in milk.
[10] Definition of subclinical ketosis: BHB ≥1.2 mmol/L in blood serum or plasma.
[11] Definition of subclinical ketosis: BHB ≥1.4 mmol/L in blood serum or plasma.

health management as well as on economics are scarce, and future research is needed.

### 6.2.4 Sensor-derived animal behavior and data integration

Combining in-line data on milk yield and its components with sensor-derived animal behaviors might enhance ketosis monitoring by identifying cows at risk of SCK that could benefit from a cow-side test (Petersson et al., 2017).

As described in Section 5.1.1, PLF technologies are used inter alia to measure animal behaviors. As hyperketonemia has an impact on neurophysiological processes, it can be assumed that it also affects movement and behavior patterns of animals as well as the time spent in certain areas of the barn, that is, their time budget.

Recently, several sensor systems have been validated to accurately characterize distinct animal behaviors, for example, rumination (Schirmann et al., 2009; Burfeind et al., 2011; Borchers et al., 2016; Reiter et al., 2018), standing and lying times (Borchers et al., 2016; Van Erp-Van der Kooj et al., 2016), and animal position (Wolfger et al., 2016). Furthermore, changes in distinct behaviors were reported to be associated with animal disease (Titler et al., 2013; Itle et al., 2015; Liboreiro et al., 2015; Stangaferro et al., 2016a,b,c).

For cows suffering from ketosis, Edwards and Tozer (2004) reported an increase in walking activity and decrease in milk yield, which were already detectable in 8 and 6 days, respectively, before clinical diagnosis by use of sensor technology (Afikim system, Kibbutz Afikim, Israel). Furthermore, the time spent feeding, dry matter intake, and social behavior were associated with the occurrence of SCK. For every 10-min decrease in average daily time spent at the feeder and for every 1 kg decrease in dry matter intake during the week before calving, the risk of SCK increased by 1.9 and 2.2 times, respectively (Goldhawk et al., 2009). The previously mentioned parameters are predisposed to capture by the use of sensor technology and, hence, can contribute to an improvement of automated monitoring of ketosis. Stangaferro et al. (2016a,b,c) published a series of papers evaluating the performance of an automated health monitoring system (HR Tags, SCR Dairy, Israel) to identify cows suffering from metabolic and digestive disorders, mastitis, and metritis. In the final dataset, 1080 cows were included, of those 52% (n = 629) had at least one health disorder. Based on rumination time and physical activity, a daily 'health index score' (HIS) between 0 and 100 was calculated for individual cows by an algorithm. Defining a positive HIS outcome as a HIS of <86 for at least 1 day in the period of interest from 5 days before to 2 days after clinical diagnosis, an SE of the system for detecting ketosis (n = 54, i.e. an incidence of 5%) of 91% was reported. In the case of ketosis, the limit of HIS <86 was reached 1.6 days (mean) earlier before clinical diagnosis by farm personnel. For cows with ketosis, rumination and activity patterns were comparable to cows in the non-diseased group for most days in the study period and both parameters did not show a dramatic decline before clinical diagnosis. This study result once again demonstrates the potential of continuous sensor-based health monitoring, in which an animal acts as its own control. For the latter, the current rumination and activity are compared with past individual animal data and is not only compared with herd members.

In a recent approach, Sturm et al. (2020) developed a flexible classification algorithm for predicting ketosis based on various input variables, consisting of time series and features from different sources. In this study, individual cow rumination data, activity measures, standing/lying behavior, and the animal's position within the barn were captured by an accelerometer system (SMARTBOW, Smartbow GmbH, Austria) from 1 week before up to 1 week after parturition. Furthermore, barn climate (temperature and humidity), visually determined body condition score, back fat thickness measured by ultrasound, and NEFAs were included in the algorithm. Animals showing a BHB concentration in coccygeal blood of >1.2 mmol/L were defined as suffering from SCK. Depending on the model, SE varied between 63% and 67%, and SP between 59% and 74%. For all models, negative predictive values were high (>91%) but should be interpreted with caution as it was highly affected by an unbalanced data structure in this study. Although the quality of the algorithm is not yet sufficient for use in commercial farms for predicting ketosis at the animal level, it is a promising approach in terms of data integration. The flexible and easily adoptable algorithm will be further developed to include additional (heterogeneous) data from other sources. In this context, merging data from different sources is the next step toward a more 'holistic' analysis of data, which aims to provide new insights into the development and prevention of diseases.

## 7 Conclusion and future trends in research

The performance achieved in today's livestock farming is the result of decades of progress, in the fields of, for example, genetics, feeding, and animal husbandry, as well as in general farm and herd health management. In addition to profound technical knowledge, farmers increasingly need to possess managerial qualities, which are necessary to ensure the long-term economic success of their farms. Nowadays, for managing the complex farms, methods of (large-scale) production of other industrial sectors, such as SOPs and risk analyses, are used, and advice is sought from external consultants. In this context, veterinarians are the most important consultants for farmers in terms of securing and improving animal health and for ensuring high levels of animal welfare (Pothmann et al., 2014). Veterinary diagnoses contribute to the pool of farm data that can be used for, for example, herd (health) management. In the best case scenario, most reliable data, which can be collected in real time, are the basis for evidence-based decisions made by farmers and veterinarians.

A cornerstone of modern herd management is the routine examination of animals for the presence of disease. The findings of epidemiological research, that already subclinical forms of the disease (i.e. without externally detectable symptoms) have a detrimental impact on animal health, welfare, and performance, has led to an increased interest in the monitoring of subclinical disease. Hence, a close monitoring of dairy cows in the transition period, for example, for detecting subclinical hypocalcemia, ketosis, mastitis, and endometritis, has become a standard procedure as part of fresh cow monitoring routines on many dairy farms.

For on-farm health monitoring, a variety of independently evaluated tests and devices are available. The use of hand-held devices for metabolic monitoring of dairy cows has attracted increasing interest from farmers and veterinarians in the last decade, and the use of these devices for ketosis monitoring has already become the standard (Overton et al., 2017). Many of these hand-held devices can be used even under harsh conditions on farms. They are easy to operate and provide reliable results in a very short time. Furthermore, displayed numerical measures are easy to interpret, in particular, compared with color change–based semi-quantitative tests. The availability of reliable test results during animal examination allows prompt treatment of diseased animals, which can contribute to improved animal welfare while securing the animals' productivity.

The use of sensor technologies in agriculture is considered as the next milestone in the development of the industry sector and is sometimes also referred to as the next industrial revolution (Zambon et al., 2019). In livestock, farmers increasingly use sensor-based PLF technologies for the monitoring of animal health, welfare, and performance. Due to the ongoing further development and cost reduction of the sensors, it can be assumed that their use in dairy farming will also continue to increase. Recently, several features of commercially available sensor systems have been validated to accurately classify distinct animal behaviors, for example, rumination, standing and lying times, as well as animal activities. Because specific behavioral patterns are associated with the disease, changes in behaviors detected by sensors are used to identify animals at risk. When PLF technologies are used as the sole test to detect specific diseases, the test performance is currently not sufficient for practical use. Nevertheless, sensor technologies can contribute to improved animal health by identifying animals at risk (e.g. subclinically diseased), which afterward can be examined with more specific tests. Without sensor technologies, these animals might be overlooked by farm personnel.

When introducing new technologies in practice, the question arises as to the economic benefit of this investment. Some models for the economic evaluation of different PLF technologies have already been developed (Steeneveld et al., 2015; Dolecheck et al., 2016). The economic value depends on numerous factors, which vary from farm to farm and situation to situation, for instance, on milk price and labor costs. Furthermore, the evaluation also depends on the number of features, which are offered for one system. With an increasing number of frequently used features, the economic benefit of a system will increase. One advantage of algorithm-based systems (e.g. CowManager Sensor; SMARTBOW®; SenseHub™ Dairy) is that the introduction of new features is usually not associated with a replacement of the entire sensor system, but, for example, 'only' with adoptions in the software. The optimization of algorithms during operation can also lead to an increase in economic efficiency. Due to these possibilities for expansion and improvement, a statement on the cost-benefit ratio is always only a snapshot. Cost-benefit calculations for the use of PLF technologies are currently under review. Furthermore, intensive research activities are focusing on the identification of additional biomarkers, for example, those used for the reliable assessment of individual animal's stress or welfare. Once these have been identified, it will then be necessary to develop reliable on-farm tests or to identify feature variables, which can be measured by sensor technologies. Linking the phenotype of an animal to measurements that can be derived from PLF technologies is also in the focus of actual research.

In the past, individual aspects of dairy farming, for example, animal nutrition, housing, climate, and animal performance, were often analyzed independently from each other, without considering potential interactions. Recent research focuses on integrating heterogeneous farm and animal data, generated by various technologies. In addition to facilitating the exchange of data between systems, integrated analyses of big data using, for example, artificial intelligence, enables new insights into the development of diseases as well as into the interaction of diseases among each other or with environmental factors, respectively. In this more 'holistic' approach, concepts of computer science, biostatistics, behavioral research, economics, animal sciences, and product development needs to interact across disciplines to gain additional information. For this, a close and trustworthy cooperation of, for example, engineering, animal science, veterinary medicine, economics, and social sciences is needed.

Tools that can also be used by laypersons in veterinary medicine, for example, farmers, are regularly criticized by some members of the veterinary profession, sometimes with the fear that they might give up a field of activity. For example, using ketometers or the automated and continuous monitoring of activity patterns by sensor technology are a tool, which not only are primarily used by farmers but also provides veterinarians with valuable information on animal and herd level. One of the advantages for veterinarians is that the identification of conspicuous animals does not have to be based solely on the farmer's skills to detect diseased animals or on single animal tests. Hence, sensor technologies have the potential of paying more attention to an individual animal within a herd.

With their expertise, veterinarians will remain a central element in herd health management, provided they are open-minded toward these technologies. For the veterinarian of the future, understanding the principles of data collection as well as their interpretation and use will be a self-evident part of the profession, just like today (and also in the future) using a stethoscope and rectal glove.

Despite all the optimism regarding the application of complex mathematical methods and the use of artificial intelligence to identify risk factors and for preventing diseases, there is still a need to maintain and sharpen our 'common sense' in animal husbandry. The potential of PLF and data integration should not be ignored by the veterinary profession and other stakeholders, as they will become valuable tools in daily practice. As a consequence of the general technical development, they are innovative tools in herd health management – nothing more, but nothing less either!

## 8 Where to look for further information

A comprehensive review of the evolution in the use of metabolic indicators for herd health management is found in 'A 100-year review: Metabolic health indicators and management of dairy cattle' presented by Overton et al. (2017).

Numerous recent publications are available on the development and use of sensor technologies in livestock farming:

- A general introduction and overview of biosensors are presented by Grieshaber et al. (2008).
- A structured review of 126 publications describing 139 sensor systems used for herd health management is provided in 'Invited review: Sensors to support health management on dairy farms' by Rutten et al. (2013).
- A comprehensive review of recent developments in the field of biosensors, including the detection of infectious agents, is presented in 'Recent advances in wearable sensors for animal health management' by Neethirajan (2017).
- An example of using artificial intelligence for the early detection of mastitis is presented by Hyde et al. (2020) in 'Automated prediction of mastitis infection patterns in dairy herds using machine learning'.

Journals and regular conferences providing recent research include:

- The *Journal of Dairy Science* is a renowned peer-reviewed dairy research journal.
- The European Association on Precision Livestock Farming (EA-PLF) holds regular meetings on PLF.
- The European Conference on Precision Livestock Farming (EC-PLF) presents innovations in PLF every 2 years.
- The World Buiatrics Congress (WBC) promotes knowledge transfer within the international veterinary community.

There are a number of current research projects on the use of modern management tools to support farmers and veterinarians in herd health management, including:

- EU-PLF focused on delivering a validated Blueprint for an animal and farm-centric approach to innovative livestock farming in Europe (www.eu -plf.eu).
- GENTORE will develop innovative genome-enabled selection and management tools to empower farmers to optimize cattle resilience and efficiency in different and changing environments (www.gentore.eu).

## 9 References

Abuelo, Á. and Alves-Nores, V. (2016). Point-of-care testing in cattle practice: reliability of cow-side diagnostic tests. *In Practice* 38(6): 293–302.

Adler, J. H., Roberts, S. J. and Steel, R. G. (1957). The relation between reactions to the Ross test on milk and urine and the degree of ketonemia in dairy cows. *The Cornell Veterinarian* 47(1): 101–111.

Afikim (2019). Afimilk's in-line milk lab and parlor automation solutions. In: Afimilk, K. A. (Ed). Israel: Kibbutz Afimilk.

Andersson, L. (1988). Subclinical ketosis in dairy cows. *Veterinary Clinics of North America: Food Animal Practice* 4(2): 233–251.

Arnold, J. E., Camus, M. S., Freeman, K. P., Giori, L., Hooijberg, E. H., Jeffery, U., Korchia, J., Meindel, M. J., Moore, A. R., Sisson, S. C., Vap, L. M. and Cook, J. R. (2019). ASVCP Guidelines: Principles of quality assurance and standards for veterinary clinical pathology (version 3.0): developed by the American Society for Veterinary Clinical Pathology's (ASVCP) Quality Assurance and Laboratory Standards (QALS) Committee. *Veterinary Clinical Pathology* 48: 542–618.

Bach, K. D., Heuwieser, W. and McArt, J. A. A. (2016). Technical note: Comparison of 4 electronic handheld meters for diagnosing hyperketonemia in dairy cows. *Journal of Dairy Science* 99: 9136–9142.

Banhazi, T. M., Lehr, H., Black, J. L., Crabtree, H., Schofield, P., Tscharke, M. and Berckmans, D. (2012). Precision Livestock Farming: an international review of scientific and commercial aspects. *International Journal of Agricultural and Biological Engineering* 5.

Banhazi, T. M., Rutley, D. L. and Pitchford, W. S. (2010). Validation and fine-tuning of a predictive model for air quality in livestock buildings. *Biosystems Engineering* 105(3): 395–401.

Bartens, M. C., Drillich, M., Rychli, K., Iwersen, M., Arnholdt, T., Meyer, L. and Klein-Jöbstl, D. (2016). Assessment of different methods to estimate bovine colostrum quality on farm. *New Zealand Veterinary Journal* 64(5): 263–267.

Berckmans, D. (2008). Precision livestock farming (PLF). *Computers and Electronics in Agriculture* 62(1): 1.

Berckmans, D. (2014). Precision livestock farming technologies for welfare management in intensive livestock systems. *Revue Scientifique et Technique* 33(1): 189–196.

Berckmans, D. and Guarino, M. (2008) Preface. *Computers and Electronics in Agriculture* 64(1): 1.

Bergman, E. N. (1971). Hyperketonemia-ketogenesis and ketone body metabolism. *Journal of Dairy Science* 54(6): 936–948.

Bertani, F., Mayfield, M. J., Mascilongo, V. and Nicols, S. (2019). Veterinary borderline products, medical devices and in vitro diagnostics: Global regulatory overview. *Regulatory Focus.* (accessed 21.04.2020).

Boivin, X., Lensink, J., Tallet, C. and Veissier, I. (2003). Stockmanship and farm animal welfare. *Animal Welfare* 12: 479–492.

Borchers, M. R. (2015). An evaluation of precision dairy farming technology adoption, perception, effectiveness, and use. Theses and Dissertations - Animal and Food Sciences, University of Kentucky.

Borchers, M. R., Chang, Y. M., Tsai, I. C., Wadsworth, B. A. and Bewley, J. M. (2016). A validation of technologies monitoring dairy cow feeding, ruminating, and lying behaviors. *Journal of Dairy Science* 99(9): 7458–7466.

Bradley, A. J. and Green, M. J. (2004). The importance of the nonlactating period in the epidemiology of intramammary infection and strategies for prevention. *Veterinary Clinics of North America: Food Animal Practice* 20(3): 547–568.

Bruss, M. L. (1989). Ketogenesis and ketosis. In: Kaneko, J. (Ed.) *Clinical Biochemistry of Domestic Animals* (4th edn.). Toronto: Academic Press, 86–105.

Burfeind, O., Schirmann, K., von Keyserlingk, M. A., Veira, D. M., Weary, D. M. and Heuwieser, W. (2011). Evaluation of a system for monitoring rumination in heifers and calves. *Journal of Dairy Science* 94(1): 426–430.

Burfeind, O., von Keyserlingk, M. A. G., Weary, D. M., Veira, D. M. and Heuwieser, W. (2010). Short communication: repeatability of measures of rectal temperature in dairy cows. *Journal of Dairy Science* 93(2): 624–627.

Cameron, R. E., Dyk, P. B., Herdt, T. H., Kaneene, J. B., Miller, R., Bucholtz, H. F., Liesman, J. S., Vandehaar, M. J. and Emery, R. S. (1998). Dry cow diet, management, and energy balance as risk factors for displaced abomasum in high producing dairy herds. *Journal of Dairy Science* 81(1): 132–139.

Carrier, J., Stewart, S., Godden, S., Fetrow, J. and Rapnicki, P. (2004). Evaluation and use of three cowside tests for detection of subclinical ketosis in early postpartum cows. *Journal of Dairy Science* 87(11): 3725–3735.

Chapinal, N., Carson, M., Duffield, T. F., Capel, M., Godden, S., Overton, M., Santos, J. E. and LeBlanc, S. J. (2011). The association of serum metabolites with clinical disease during the transition period. *Journal of Dairy Science* 94(10): 4897–4903.

Chapinal, N., LeBlanc, S. J., Carson, M. E., Leslie, K. E., Godden, S., Capel, M., Santos, J. E., Overton, M. W. and Duffield, T. F. (2012). Herd-level association of serum metabolites in the transition period with disease, milk production, and early lactation reproductive performance. *Journal of Dairy Science* 95(10): 5676-5682.

Clark Jr., L. C. and Lyons, C. (1962). Electrode systems for continuous monitoring in cardiovascular surgery. *Annals of the New York Academy of Sciences* 102: 29-45.

Cook, N., Oetzel, G. and Nordlund, K. (2006a). Modern techniques for monitoring high-producing dairy cows - 1. Principles of herd-level diagnoses. *In Practice* 28(9): 510-515.

Cook, N., Oetzel, G. and Nordlund, K. (2006b). Modern techniques for monitoring high-producing dairy cows - 2. Practical applications. *In Practice* 28(10): 598-603.

CowManager. Available at: https://www.cowmanager.com/en-us/ (accessed 29 December 2020).

Curtis, C. R., Erb, H. N., Sniffen, C. J., Smith, R. D. and Kronfeld, D. S. (1985). Path analysis of dry period nutrition, postpartum metabolic and reproductive disorders, and mastitis in Holstein cows. *Journal of Dairy Science* 68(9): 2347-2360.

Custer, E. M., Myers, J. L., Poffenbarger, P. L. and Schoen, I. (1983). The storage stability of 3-hydroxybutyrate in serum, plasma, and whole blood. *American Journal of Clinical Pathology* 80(3): 375-380.

Da Silva, J. C., Noordhuizen, J. P. T. M., Vagneur, M., Bexiga, R., Gelfert, C. C. and Baumgartner, W. (2006). Veterinary dairy herd health management in Europe: constraints and perspectives. *Veterinary Quarterly* 28(1): 23-32.

De Koning, C. (2010). Automatic milking - common practice on dairy farms. Proceedings of the 2nd North American Conference on Robotic Milking. Toronto, Canada: Precision Dairy Operators, V59-V63.

De Roos, A. P. W., Van Den Bijgaart, H. J. C. M., Hørlyk, J. and de Jong, G. (2007). Screening for subclinical ketosis in dairy cattle by Fourier transform infrared spectrometry. *Journal of Dairy Science* 90(4): 1761-1766.

Denis-Robichaud, J., Dubuc, J., Lefebvre, D. and DesCôteaux, L. (2014). Accuracy of milk ketone bodies from flow-injection analysis for the diagnosis of hyperketonemia in dairy cows. *Journal of Dairy Science* 97(6): 3364-3370.

Detilleux, J. C., Kehrli Jr., M. E., Stabel, J. R., Freeman, A. E. and Kelley, D. H. (1995). Study of immunological dysfunction in periparturient Holstein cattle selected for high and average milk production. *Veterinary Immunology and Immunopathology* 44(3-4): 251-267.

Dohoo, I. R. and Martin, S. W. (1984). Subclinical ketosis: prevalence and associations with production and disease. *Canadian Journal of Comparative Medicine (Revue Canadienne de Medecine Comparee)* 48(1): 1-5.

Dohoo, I. R., Martin, S. W. and Stryhn, H. (2009). *Veterinary Epidemiologic Research*. Charlottetown, PE, Canada: VER Incorporated.

Dolecheck, K. A., Heersche Jr., G. and Bewley, J. M. (2016). Retention payoff-based cost per day open regression equations: application in a user-friendly decision support tool for investment analysis of automated estrus detection technologies. *Journal of Dairy Science* 99(12): 10182-10193.

Drackley, J. K. (1999). ADSA Foundation Scholar Award. Biology of dairy cows during the transition period: the final frontier? *Journal of Dairy Science* 82(11): 2259-2273.

Drillich, M. and Wagener, K. (2018). Pathogenesis of uterine diseases in dairy cattle and implications for fertility. *Animal Reproduction* 15(Suppl. 1): 879-885.

Duffield, T. F. (1997). Effects of a monensin controlled release capsule on energy metabolism, health and production in lactating dairy cattle. Guelph, Canada: University of Guelph.

Duffield, T. (2000). Subclinical ketosis in lactating dairy cattle. *Veterinary Clinics of North America: Food Animal Practice* 16(2): 231–253.

Duffield, T. F., Kelton, D. F., Leslie, K. E., Lissemore, K. D. and Lumsden, J. H. (1997). Use of test day milk fat and milk protein to detect subclinical ketosis in dairy cattle in Ontario. *Canadian Veterinary Journal = la Revue Veterinaire Canadienne* 38(11): 713–718.

Duffield, T. F., Lissemore, K. D., McBride, B. W. and Leslie, K. E. (2009). Impact of hyperketonemia in early lactation dairy cows on health and production. *Journal of Dairy Science* 92(2): 571–580.

Eastwood, C. R., Chapman, D. F. and Paine, M. S. (2012). Networks of practice for co-construction of agricultural decision support systems: case studies of precision dairy farms in Australia. *Agricultural Systems* 108: 10–18.

Edwards, J. L. and Tozer, P. R. (2004). Using activity and milk yield as predictors of fresh cow disorders. *Journal of Dairy Science* 87(2): 524–531.

Egger-Danner, C., Fuerst-Waltl, B., Obritzhauser, W., Fuerst, C., Schwarzenbacher, H., Grassauer, B., Mayerhofer, M. and Koeck, A. (2012). Recording of direct health traits in Austria – Experience report with emphasis on aspects of availability for breeding purposes. *Journal of Dairy Science* 95(5): 2765–2777.

Erb, H. N. and Gröhn, Y. T. (1988). Epidemiology of metabolic disorders in the periparturient dairy cow. *Journal of Dairy Science* 71(9): 2557–2571.

Espadamala, A., Pallarés, P., Lago, A. and Silva-Del-Río, N. (2016). Fresh-cow handling practices and methods for identification of health disorders on 45 dairy farms in California. *Journal of Dairy Science* 99(11): 9319–9333.

Fetrow, J., McClary, D., Harman, R., Butcher, K., Weaver, L., Studer, E., Ehrlich, J., Etherington, W., Guterbock, W., Klingborg, D., Reneau, J. and Williamson, N. (1990). Calculating selected reproductive indices: recommendations of the American Association of Bovine Practitioners. *Journal of Dairy Science* 73(1): 78–90.

Fetrow, J., Nordlund, K. V. and Norman, H. D. (2006). Invited review: culling: nomenclature, definitions, and recommendations. *Journal of Dairy Science* 89(6): 1896–1905.

Fiore, F., Cocco, R., Musina, D. and Spissu, N. (2020). On-farm use of a water hardness test kit to assess total blood calcium level in dairy cattle. *Journal of Dairy Research* 87(1): 56–59.

Flatland, B., Freeman, K. P., Vap, L. M., Harr, K. E. and ASVCP (2013). ASVCP guidelines: quality assurance for point-of-care testing in veterinary medicine. *Veterinary Clinical Pathology* 42(4): 405–423.

Forshell, K. P. and Østerås, O. (2001). Trends in Norwegian udder health data during 1975 through 2000. Proceedings of the 2nd International Symposium on Mastitis and Milk Quality, Madison, WI: National Mastitis Council.

Galen, R. S. and Gambino, S. R. (1975). *Beyond Normality: The Predictive Value and Efficiency of Medical Diagnoses*. New Jersey, USA: John Wiley & Sons.

Gattani, A., Singh, S. V., Agrawal, A., Khan, M. H. and Singh, P. (2019). Recent progress in electrochemical biosensors as point of care diagnostics in livestock health. *Analytical Biochemistry* 579: 25–34.

Geishauser, T., Leslie, K., Kelton, D. and Duffield, T. (1998). Evaluation of five cowside tests for use with milk to detect subclinical ketosis in dairy cows. *Journal of Dairy Science* 81(2): 438–443.

Geishauser, T., Leslie, K., Tenhag, J. and Bashiri, A. (2000). Evaluation of eight cow-side ketone tests in milk for detection of subclinical ketosis in dairy cows. *Journal of Dairy Science* 83(2): 296-299.

George, J. W. (2001). The usefulness and limitations of hand-held refractometers in veterinary laboratory medicine: an historical and technical review. *Veterinary Clinical Pathology* 30(4): 201-210.

Godden, S. (2008). Colostrum management for dairy calves. *Veterinary Clinics of North America: Food Animal Practice* 24(1): 19-39.

Goldhawk, C., Chapinal, N., Veira, D. M., Weary, D. M. and von Keyserlingk, M. A. (2009). Prepartum feeding behavior is an early indicator of subclinical ketosis. *Journal of Dairy Science* 92(10): 4971-4977.

Grieshaber, D., MacKenzie, R., Vörös, J. and Reimhult, E. (2008). Electrochemical biosensors – sensor principles and architectures. *Sensors* 8(3): 1400-1458.

Grieve, D. G., Korver, S., Rijpkema, Y. S. and Hof, G. (1986). Relationship between milk composition and some nutritional parameters in early lactation. *Livestock Production Science* 14(3): 239-254.

Grummer, R. R. (1995). Impact of changes in organic nutrient metabolism on feeding the transition dairy cow. *Journal of Animal Science* 73(9): 2820-2833.

Grummer, R. R. (2008). Nutritional and management strategies for the prevention of fatty liver in dairy cattle. *The Veterinary Journal* 176(1): 10-20.

Grummer, R. R., Mashek, D. G. and Hayirli, A. (2004). Dry matter intake and energy balance in the transition period. *Veterinary Clinics of North America. Food Animal Practice* 20(3): 447-470.

Grummer, R. R. and Rastani, R. R. (2003). When should lactating cows reach positive energy balance? *The Professional Animal Scientist* 19(3): 197-203.

Guerci, B., Tubiana-Rufi, N., Bauduceau, B., Bresson, R., Cuperlier, A., Delcroix, C., Durain, D., Fermon, C., Le Floch, J. P., Le Devehat, C., Melki, V., Monnier, L., Mosnier-Pudar, H., Taboulet, P. and Hanaire-Broutin, H. (2005). Advantages to using capillary blood β-hydroxybutyrate determination for the detection and treatment of diabetic ketosis. *Diabetes and Metabolism* 31(4 Pt 1): 401-406.

Gussmann, M., Steeneveld, W., Kirkeby, C., Hogeveen, H., Farre, M. and Halasa, T. (2019). Economic and epidemiological impact of different intervention strategies for subclinical and clinical mastitis. *Preventive Veterinary Medicine* 166: 78-85.

Gusterer, E., Lidauer, L., Berger, A., Kickinger, F., Öhlschuster, M., Auer, W., Drillich, M. and Iwersen, M. (2019). Using the SMARTBOW system for monitoring animals suffering from periparturient disease. Proceedings of the 9th European Conference on Precision Livestock Farming, ECPLF 2019, 773-778.

Guterbock, W. M. (2004). Diagnosis and treatment programs for fresh cows. *Veterinary Clinics of North America: Food Animal Practice* 20(3): 605-626.

Hamann, J. and Krömker, V. (1997). Potential of specific milk composition variables for cow health management. *Livestock Production Science* 48(3): 201-208.

Hansen, P. W. (1999). Screening of dairy cows for ketosis by use of infrared spectroscopy and multivariate calibration. *Journal of Dairy Science* 82(9): 2005-2010.

Herdt, T. H. (2000). Ruminant adaptation to negative energy balance. Influences on the etiology of ketosis and fatty liver. *Veterinary Clinics of North America: Food Animal Practice* 16(2): 215-230.

Heuer, C., Schukken, Y. H. and Dobbelaar, P. (1999). Postpartum body condition score and results from the first test day milk as predictors of disease, fertility, yield, and culling in commercial dairy herds. *Journal of Dairy Science* 82(2): 295–304.

Heuwieser, W., Iwersen, M., Gossellin, J. and Drillich, M. (2010). Short communication: survey of fresh cow management practices of dairy cattle on small and large commercial farms. *Journal of Dairy Science* 93(3): 1065–1068.

Hyde, R. M., Down, P. M., Bradley, A. J., Breen, J. E., Hudson, C., Leach, K. A. and Green, M. J. (2020). Automated prediction of mastitis infection patterns in dairy herds using machine learning. *Scientific Reports* 10(1): 4289.

Ingvartsen, K. L. (2006). Feeding- and management-related diseases in the transition cow: physiological adaptations around calving and strategies to reduce feeding-related diseases. *Animal Feed Science and Technology* 126(3–4): 175–213.

Ingvartsen, K. L. and Andersen, J. B. (2000). Integration of metabolism and intake regulation: a review focusing on periparturient animals. *Journal of Dairy Science* 83(7): 1573–1597.

Itle, A. J., Huzzey, J. M., Weary, D. M. and von Keyserlingk, M. A. (2015). Clinical ketosis and standing behavior in transition cows. *Journal of Dairy Science* 98(1): 128–134.

Iwersen, M., Falkenberg, U., Voigtsberger, R., Forderung, D. and Heuwieser, W. (2009). Evaluation of an electronic cowside test to detect subclinical ketosis in dairy cows. *Journal of Dairy Science* 92(6): 2618–2624.

Iwersen, M., Klein-Jöbstl, D., Pichler, M., Roland, L., Fidlschuster, B., Schwendenwein, I. and Drillich, M. (2013). Comparison of 2 electronic cowside tests to detect subclinical ketosis in dairy cows and the influence of the temperature and type of blood sample on the test results. *Journal of Dairy Science* 96(12): 7719–7730.

Jenkins, N. T., Peña, G., Risco, C., Barbosa, C. C., Vieira-Neto, A. and Galvão, K. N. (2015). Utility of inline milk fat and protein ratio to diagnose subclinical ketosis and to assign propylene glycol treatment in lactating dairy cows. *Canadian Veterinary Journal = la Revue Veterinaire Canadienne* 56(8): 850–854.

Jeppesen, R., Enemark, J. M. and Enevoldsen, C. (2006). Ketone body measurement in dairy cows. Proceedings of the 24th World Buiatrics Congress, Nice, France. Vienna, Austria: World Association Buiatrics, OS 43-42.

Jones, G., Bork, O., Ferguson, S. A. and Bates, A. (2019). Comparison of an on-farm point-of-care diagnostic with conventional culture in analysing bovine mastitis samples. *Journal of Dairy Research* 86(2): 222–225.

Kanz, P., Drillich, M., Klein-Jöbstl, D., Mair, B., Borchardt, S., Meyer, L., Schwendenwein, I. and Iwersen, M. (2015). Suitability of capillary blood obtained by a minimally invasive lancet technique to detect subclinical ketosis in dairy cows by using 3 different electronic hand-held devices. *Journal of Dairy Science* 98(9): 6108–6118.

Khol, J. L., Freigassner, K., Stanitznig, A., Tichy, A. and Wittek, T. (2019). Evaluation of a handheld device for the measurement of beta-hydroxybutyrate in capillary blood obtained by the puncture of the vulva as well as in venous whole blood in cattle. *Polish Journal of Veterinary Sciences* 22: 557–564.

King, M. T. M., Duffield, T. F. and DeVries, T. J. (2019). Short communication: assessing the accuracy of inline milk fat-to-protein ratio data as an indicator of hyperketonemia in dairy cows in herds with automated milking systems. *Journal of Dairy Science* 102(9): 8417–8422.

Klein, D., Kern, A., Lapan, G., Benetka, V., Möstl, K., Hassl, A. and Baumgartner, W. (2009). Evaluation of rapid assays for the detection of bovine coronavirus, Rotavirus A and

*Cryptosporidium parvum* in faecal samples of calves. *Veterinary Journal* 182(3): 484–486.

Koenraadt, C. J. M., Balenghien, T., Carpenter, S., Ducheyne, E., Elbers, A. R., Fife, M., Garros, C., Ibáñez-Justicia, A., Kampen, H., Kormelink, R. J., Losson, B., van der Poel, W. H., De Regge, N., van Rijn, P. A., Sanders, C., Schaffner, F., Sloet van Oldruitenborgh-Oosterbaan, M. M., Takken, W., Werner, D. and Seelig, F. (2014). Bluetongue, Schmallenberg – what is next? Culicoides-borne viral diseases in the 21st century. *BMC Veterinary Research* 10: 77.

Lacasse, P., Vanacker, N., Ollier, S. and Ster, C. (2018). Innovative dairy cow management to improve resistance to metabolic and infectious diseases during the transition period. *Research in Veterinary Science* 116: 40–46.

Lago, A. and Godden, S. M. (2018). Use of rapid culture systems to guide clinical mastitis treatment decisions. *Veterinary Clinics of North America - Food Animal Practice* 34(3): 389–412.

Leal Yepes, F. A., Nydam, D. V., Heuwieser, W. and Mann, S. (2018). Technical note: evaluation of the diagnostic accuracy of 2 point-of-care β-hydroxybutyrate devices in stored bovine plasma at room temperature and at 37°C. *Journal of Dairy Science* 101(7): 6455–6461.

LeBlanc, S. J. (2006). *Monitoring Programs for Transition Dairy Cows. XXIV*. Nice, France: World Buiatrics Congress, 13.

LeBlanc, S. J. (2010). Monitoring metabolic health of dairy cattle in the transition period. *Journal of Reproduction and Development* 56 (Suppl.): 29–35.

LeBlanc, S. J., Leslie, K. E. and Duffield, T. F. (2005). Metabolic predictors of displaced abomasum in dairy cattle. *Journal of Dairy Science* 88(1): 159–170.

LeBlanc, S. J., Lissemore, K. D., Kelton, D. F., Duffield, T. F. and Leslie, K. E. (2006). Major advances in disease prevention in dairy cattle. *Journal of Dairy Science* 89(4): 1267–1279.

Liboreiro, D. N., Machado, K. S., Silva, P. R. B., Maturana, M. M., Nishimura, T. K., Brandão, A. P., Endres, M. I. and Chebel, R. C. (2015). Characterization of peripartum rumination and activity of cows diagnosed with metabolic and uterine diseases. *Journal of Dairy Science* 98(10): 6812–6827.

Lichtmannsperger, K., Hinney, B., Joachim, A. and Wittek, T. (2019). Molecular characterization of *Giardia intestinalis* and *Cryptosporidium parvum* from calves with diarrhoea in Austria and evaluation of point-of-care tests. *Comparative Immunology, Microbiology and Infectious Diseases* 66: 101333.

Luginbühl, A., Reitt, K., Metzler, A., Kollbrunner, M., Corboz, L. and Deplazes, P. (2005). Field study about prevalence and diagnostics of diarrhea causing agents in the new-born calf in a Swiss veterinary practice area. *Schweizer Archiv für Tierheilkunde* 147(6): 245–252.

Lumsden, J. H., Mullen, K. and Rowe, R. (1980). Hematology and biochemistry reference values for female Holstein cattle. *Canadian Journal of Comparative Medicine : Revue Canadienne de Medecine Comparee* 44(1): 24–31.

Macmillan, K., López Helguera, I., Behrouzi, A., Gobikrushanth, M., Hoff, B. and Colazo, M. G. (2017). Accuracy of a cow-side test for the diagnosis of hyperketonemia and hypoglycemia in lactating dairy cows. *Research in Veterinary Science* 115: 327–331.

Madoz, L. V., Prunner, I., Jaureguiberry, M., Gelfert, C. C., de la Sota, R. L., Giuliodori, M. J. and Drillich, M. (2017). Application of a bacteriological on-farm test to reduce antimicrobial usage in dairy cows with purulent vaginal discharge. *Journal of Dairy Science* 100(5): 3875-3882.

Mahrt, A., Burfeind, O. and Heuwieser, W. (2014a). Effects of time and sampling location on concentrations of β-hydroxybutyric acid in dairy cows. *Journal of Dairy Science* 97(1): 291-298.

Mahrt, A., Burfeind, O. Voigtsberger, R., Müller, A. and Heuwieser, W. (2014b). Evaluation of a new electronic handheld meter for measurement of β-hydroxybutyric acid in dairy cows. *Tierärztliche Praxis Großtiere* 42(1): 5-10.

Matsas, D. J., Warnick, L. D., Mechor, G. D., Seib, L. N., Fatone, S., White, M. E. and Guard, C. L. (1999). Use of a water hardness test kit to measure serum calcium concentration in cattle. *Journal of the American Veterinary Medical Association* 214(6): 826-828.

Mazeris, F. (2010). DeLaval Herd Navigator: proactive herd management. Proceedings of the First North American Conference on Precision Dairy Management. Toronto, Canada, 26-27.

McArt, J. A. A., Nydam, D. V. and Oetzel, G. R. (2012). A field trial on the effect of propylene glycol on displaced abomasum, removal from herd, and reproduction in fresh cows diagnosed with subclinical ketosis. *Journal of Dairy Science* 95(5): 2505-2512.

McGuirk, S. M. and Collins, M. (2004). Managing the production, storage, and delivery of colostrum. *Veterinary Clinics of North America: Food Animal Practice* 20(3): 593-603.

McKenna, S. L. B. and Dohoo, I. R. (2006). Using and interpreting diagnostic tests. *Veterinary Clinics of North America: Food Animal Practice* 22(1): 195-205.

Miettinen, P. V. A. and Setala, J. J. (1993). Relationships between subclinical ketosis, milk production and fertility in Finnish dairy cattle. *Preventive Veterinary Medicine* 17(1-2): 1-8.

Müller, R. and Schrader, L. (2003). A new method to measure behavioural activity levels in dairy cows. *Applied Animal Behaviour Science* 83(4): 247-258.

Neethirajan, S. (2017). Recent advances in wearable sensors for animal health management. *Sensing and Bio-Sensing Research* 12: 15-29.

Negussie, M., Lidauer, M., Mäntysaari, E. A., Stranden, I., Pösö, J., Nielsen, U. S., Johansson, K., Eriksson, J. A. and Aamand, G. P. (2010). Combining test day SCS with clinical mastitis and udder type traits: a random regression model for joint genetic evaluation of udder health in Denmark, Finland and Sweden. *Interbull Bulletin* 42: 25-31.

Nelson, A. J. and Redlus, H. W. (1989). The key role of records in a production medicine practice. *Veterinary Clinics of North America: Food Animal Practice* 5(3): 517-552.

Nielen, M., Aarts, M. G., Jonkers, A. G., Wensing, T. and Schukken, Y. H. (1994). Evaluation of two cowside tests for the detection of subclinical ketosis in dairy cows. *Canadian Veterinary Journal = la Revue Veterinaire Canadienne* 35(4): 229-232.

Nielsen, N. I., Friggens, N. C., Chagunda, M. G. and Ingvartsen, K. L. (2005). Predicting risk of ketosis in dairy cows using in-line measurements of beta-hydroxybutyrate: a biological model. *Journal of Dairy Science* 88(7): 2441-2453.

Nielsen, U. S., Aamand, G. P. and Mark, T. (2000). National genetic evaluation of udder health and other health traits in Denmark. *Interbull Bulletin* 25: 143-150.

Nonnecke, B. J., Kimura, K., Goff, J. P. and Kehrli, M. E. (2003). Effects of the mammary gland on functional capacities of blood mononuclear leukocyte populations from periparturient cows. *Journal of Dairy Science* 86(7): 2359-2368.

Oetzel, G. R. (2004). Monitoring and testing dairy herds for metabolic disease. *Veterinary Clinics of North America: Food Animal Practice* 20(3): 651–674.

Osborne, T. M., Leslie, K. E., Duffield, T., Petersson, C. S., TenHag, J. and Okada, Y. (2002). Evaluation of Keto-Test in urine and milk for the detection of subclinical ketosis in periparturient Holstein dairy cattle. Proceedings of the 35th Annual Conference of the American Association of Bovine Practitioners. Madison, WI, 188–189.

Ospina, P. A., Nydam, D. V., Stokol, T. and Overton, T. R. (2010). Association between the proportion of sampled transition cows with increased nonesterified fatty acids and [beta]-hydroxybutyrate and disease incidence, pregnancy rate, and milk production at the herd level. *Journal of Dairy Science* 93(8): 3595–3601.

Overton, T. R., McArt, J. A. A. and Nydam, D. V. (2017). A 100-year review: metabolic health indicators and management of dairy cattle. *Journal of Dairy Science* 100(12): 10398–10417.

Pannwitz, G. (2015). Standardized analysis of German cattle mortality using national register data. *Preventive Veterinary Medicine* 118(4): 260–270.

Panousis, N., Kritsepi, M., Karagiannis, I., Kalaitzakis, E., Lafi, S. and Brozos, C. (2011). Evaluation of Precision Xceed® for on-site monitoring of blood β-hydroxybutyric acid and glucose in dairy cows. *Journal of the Hellenic Veterinary Medical Society* 62: 109–117.

Petersson, C. S., Steele, N. M., Swartz, T. H. and Dela Rue, B. T. (2017). Opportunities for identifying animal health and well-being disorders using precision technologies. In: Beede, D. K. (Ed.) *Large Dairy Herd Management* (3rd edn.). Champaign, IL: American Dairy Science Association, 1279–1292.

Philipsson, J. and Lindhé, B. (2003). Experiences of including reproduction and health traits in Scandinavian dairy cattle breeding programmes. *Livestock Production Science* 83(2–3): 99–112.

Pineda, A. and Cardoso, F. C. (2015). Technical note: Validation of a handheld meter for measuring β-hydroxybutyrate concentrations in plasma and serum from dairy cows. *Journal of Dairy Science* 98(12): 8818–8824.

Pothmann, H., Nechanitzky, K., Sturmlechner, F. and Drillich, M. (2014). Consultancy to dairy farmers relating to animal health and herd health management on small- and medium-sized farms. *Journal of Dairy Science* 97(2): 851–860.

Pralle, R. S. and White, H. M. (2020) Symposium review: Big data, big predictions: utilizing milk Fourier-transform infrared and genomics to improve hyperketonemia management. *Journal of Dairy Science* 103(4): 3867–3873.

Pyörälä, S. (2008). Mastitis in post-partum dairy cows. *Reproduction in Domestic Animals = Zuchthygiene* 43 (Suppl. 2): 252–259.

Raboisson, D., Mounié, M. and Maigné, E. (2014). Diseases, reproductive performance, and changes in milk production associated with subclinical ketosis in dairy cows: a meta-analysis and review. *Journal of Dairy Science* 97(12): 7547–7563.

Reiter, S., Sattlecker, G., Lidauer, L., Kickinger, F., Öhlschuster, M., Auer, W., Schweinzer, V., Klein-Jöbstl, D., Drillich, M. and Iwersen, M. (2018). Evaluation of an ear-tag-based accelerometer for monitoring rumination in dairy cows. *Journal of Dairy Science* 101(4): 3398–3411.

Reneau, J. K. (1986). Effective use of Dairy Herd Improvement somatic cell counts in mastitis control. *Journal of Dairy Science* 69(6): 1708–1720.

Roberts, T., Chapinal, N., LeBlanc, S. J., Kelton, D. F., Dubuc, J. and Duffield, T. F. (2012). Metabolic parameters in transition cows as indicators for early-lactation culling risk. *Journal of Dairy Science* 95(6): 3057-3063.

Rodriguez, E. M., Aris, A. and Bach, A. (2017). Associations between subclinical hypocalcemia and postparturient diseases in dairy cows. *Journal of Dairy Science* 100(9): 7427-7434.

Rollin, F. (2006). Tool for a promt cowside diagnosis: what can be implemented by the bovine practitioner? Proceedings of the 24th World Buiatrics Congress, Nice, France. Vienna, Austria: World Association for Buiatrics.

Rothera, A. C. H. (1908). Note on the sodium nitro-prusside reaction for acetone. *The Journal of Physiology* 37(5-6): 491-494.

Rushen, J., de Passille, A. M., von Keyserlingk, M. A. G. and Weary, D. M. (2008). *The Welfare of Cattle*. The Netherlands: Springer.

Rutten, C. J., Velthuis, A. G. J., Steeneveld, W. and Hogeveen, H. (2013). Invited review: Sensors to support health management on dairy farms. *Journal of Dairy Science* 96(4): 1928-1952.

Schalm, O. W. and Noorlander, D. O. (1957). Experiments and observations leading to development of the *California mastitis* test. *Journal of the American Veterinary Medical Association* 130(5): 199-204.

Schirmann, K., von Keyserlingk, M. A. G., Weary, D. M., Veira, D. M. and Heuwieser, W. (2009). Technical note: Validation of a system for monitoring rumination in dairy cows. *Journal of Dairy Science* 92(12): 6052-6055.

SenseHub. Available at: https://www.allflex.global/product/sensehub-dairy/ (accessed 29 December 2020).

Sheldon, I. M., Molinari, P. C. C., Ormsby, T. J. R. and Bromfield, J. J. (2020). Preventing postpartum uterine disease in dairy cattle depends on avoiding, tolerating and resisting pathogenic bacteria. *Theriogenology* 150: 158-165.

SMARTBOW. Available at: www.smartbow.com.

Stangaferro, M. L., Wijma, R., Caixeta, L. S., Al-Abri, M. A. and Giordano, J. O. (2016a). Use of rumination and activity monitoring for the identification of dairy cows with health disorders: Part I. Metabolic and digestive disorders. *Journal of Dairy Science* 99(9): 7395-7410.

Stangaferro, M. L., Wijma, R., Caixeta, L. S., Al-Abri, M. A. and Giordano, J. O. (2016b). Use of rumination and activity monitoring for the identification of dairy cows with health disorders: Part II. Mastitis. *Journal of Dairy Science* 99(9): 7411-7421.

Stangaferro, M. L., Wijma, R., Caixeta, L. S., Al-Abri, M. A. and Giordano, J. O. (2016c). Use of rumination and activity monitoring for the identification of dairy cows with health disorders: Part III. Metritis. *Journal of Dairy Science* 99(9): 7422-7433.

Steeneveld, W., Hogeveen, H. and Oude Lansink, A. G. J. M. (2015). Economic consequences of investing in sensor systems on dairy farms. *Computers and Electronics in Agriculture* 119: 33-39.

Steeneveld, W., Tauer, L. W., Hogeveen, H. and Oude Lansink, A. G. (2012). Comparing technical efficiency of farms with an automatic milking system and a conventional milking system. *Journal of Dairy Science* 95(12): 7391-7398.

Stewart, S., Fetrow, J. and Eicker, S. (1994). Analysis of current performance on commercial dairies. *Compendium on Continuing Education for the Practicing Veterinarian* 16: 1099-1103.

Stokol, T. and Nydam, D. V. (2005). Effect of anticoagulant and storage conditions on bovine nonesterified fatty acid and beta-hydroxybutyrate concentrations in blood. *Journal of Dairy Science* 88(9): 3139-3144.

Sturm, V., Efrosinin, D., Öhlschuster, M., Gusterer, E., Drillich, M. and Iwersen, M. (2020). Combination of sensor data and health monitoring for early detection of subclinical Ketosis in dairy cows. *Sensors* 20(5).

Süss, D., Drillich, M., Klein-Jöbstl, D., Wagener, K., Krieger, S., Thiel, A., Meyer, L., Schwendenwein, I. and Iwersen, M. (2016). Measurement of β-hydroxybutyrate in capillary blood obtained from an ear to detect hyperketonemia in dairy cows by using an electronic handheld device. *Journal of Dairy Science* 99(9): 7362-7369.

Suthar, V. S., Canelas-Raposo, J., Deniz, A. and Heuwieser, W. (2013). Prevalence of subclinical ketosis and relationships with postpartum diseases in European dairy cows. *Journal of Dairy Science* 96(5): 2925-2938.

Titler, M., Maquivar, M. G., Bas, S., Gordon, E., Rajala-Schultz, P. J., McCullough, K. and Schuenemann, G. M. (2013). Effect of metritis on daily activity patterns in lactating Holstein dairy cows. *Journal of Dairy Science* 96(1): 647.

Tyopponen, J. and Kauppinen, K. (1980). The stability and automatic determination of ketone bodies in blood samples taken in field conditions. *Acta Veterinaria Scandinavica* 21(1): 55-61.

Van Asseldonk, M. A. P. M. (1999). *Economic Evaluation of Information Technology Applications on Dairy Farms*. S.l.: Van Asseldonk.

Van der Drift, S. G. A., Jorritsma, R., Schonewille, J. T., Knijn, H. M. and Stegeman, J. A. (2012). Routine detection of hyperketonemia in dairy cows using Fourier transform infrared spectroscopy analysis of β-hydroxybutyrate and acetone in milk in combination with test-day information. *Journal of Dairy Science* 95(9): 4886-4898.

Van Erp-Van der Kooj, E., Van de Brug, M. and Roelofs, J. (2016). Validation of the Nedap Smarttag Leg and Neck to assess behavioral activity level in dairy cattle. In: Kamphuis, C. and Steeneveld, W. (Eds) *Precision Dairy Farming 2016*. Wageningen Academic Publishers, 321-326.

Van Knegsel, A. T. M., Van der Drift, S. G. A., Horneman, M., de Roos, A. P., Kemp, B. and Graat, E. A. (2010). Short communication: ketone body concentration in milk determined by Fourier transform infrared spectroscopy: value for the detection of hyperketonemia in dairy cows. *Journal of Dairy Science* 93(7): 3065-3069.

Venjakob, P. L., Borchardt, S. and Heuwieser, W. (2017). Hypocalcemia - cow-level prevalence and preventive strategies in German dairy herds. *Journal of Dairy Science* 100(11): 9258-9266.

Villa-Godoy, A., Hughes, T. L., Emery, R. S., Chapin, L. T. and Fogwell, R. L. (1988). Association between energy balance and luteal function in lactating dairy cows. *Journal of Dairy Science* 71(4): 1063-1072.

Voyvoda, H. and Erdogan, H. (2010). Use of a hand-held meter for detecting subclinical ketosis in dairy cows. *Research in Veterinary Science* 89(3): 344-351.

Wagener, K., Gabler, C. and Drillich, M. (2017). A review of the ongoing discussion about definition, diagnosis and pathomechanism of subclinical endometritis in dairy cows. *Theriogenology* 94: 21-30.

Walsh, R. B., Walton, J. S., Kelton, D. F., LeBlanc, S. J., Leslie, K. E. and Duffield, T. F. (2007). The effect of subclinical ketosis in early lactation on reproductive performance of postpartum dairy cows. *Journal of Dairy Science* 90(6): 2788-2796.

Webster, J. (2016). Animal welfare: freedoms, dominions and 'A life worth living.' *Animals* 6(6): 35.

Wilson, A. J. and Mellor, P. S. (2009). Bluetongue in Europe: past, present and future. *Philosophical Transactions of the Royal Society of London. Series B, Biological Sciences* 364(1530): 2669-2681.

Wischenbart, M., Tomic, D., Iwersen, M., Schrefl, M. and Sturm, V. (2017). agriProKnow - process-oriented information management in precision dairy farming. In: (KTBL) AfTaSiA (Eds) 13th Conference on Contruction, Engineering and Environment in Livestock Farming. Stuttgart-Hohenheim.

Wolfger, B., Jones, B. W., Orsel, K. and Bewley, J. M. (2016). Technical note: Evaluation of an ear-attached real-time location-monitoring system. *Journal of Dairy Science* 100(3): 2219-2224.

Zakian, A., Tehrani-Sharif, M., Mokhber-Dezfouli, M. R., Nouri, M. and Constable, P. D. (2017). Evaluation of a point-of-care electrochemical meter to detect subclinical ketosis and hypoglycaemia in lactating dairy cows. *Australian Veterinary Journal* 95: 123-128.

Zambon, I., Cecchini, M., Egidi, G., Saporito, M. G. and Colantoni, A. (2019). Revolution 4.0: industry vs. agriculture in a future development for SMEs. *Processes* 7(1): 36.

# Chapter 4

## Advances in precision livestock farming techniques for monitoring dairy cattle welfare

*Henk Hogeveen and Mariska van der Voort, Wageningen University and Research, The Netherlands*

## 1 Introduction

Society is increasingly interested about animal well-being in food-producing animals (Winckler, 2019). As a consequence, more and more regulations are created by public (government) and private (partners in the supply chain) entities to ensure animal welfare. Dairy cattle welfare is no exception to this trend. Welfare assessment systems are needed in order to study animal welfare. These can be very intensive (animal observation) systems which are used for research to, for instance, assess the impact of management and farm structure (housing) on dairy cattle welfare. These studies form the basis for welfare regulations.

Besides the use for regulations, there is a demand for methods to assess the overall welfare of farm animals. First attempts to assess animal welfare were resource based, because assessing minimum resource requirements is generally easier than to evaluate the impact of the production conditions on animals (Mench, 2003). Resource-based assessment intends to warrant

http://dx.doi.org/10.19103/AS.2020.0084.06

the provision of the necessary environmental conditions for optimal animal welfare. Animal-based assessment methods that were developed, such as the Welfare Quality® assessment protocol (Botreau et al., 2007), provide a thorough assessment of the impact of the actual rearing conditions on welfare. Country-specific protocols, such as the National Dairy FARM (Farmers Assuring Responsible Management) Program in the USA and the AssureWel program in the UK, are also developed and propagated. Moreover, individual milk processing cooperatives/companies are implementing their own dairy cattle welfare programs, often in collaboration with large customers.

All of these programs are mostly resource-oriented. Farm assessment happens on paper as well as through regular (but not frequent) farm visits by an assessor. This makes the implementation of comprehensive animal welfare assessment protocols at a commercial level difficult, time-consuming and expensive. Precision livestock farming (PLF) can offer tools that enable a farmer to monitor animals automatically, and thus to also monitor their health and welfare (Berckmans, 2014; Halachmi et al., 2019). Sensors (e.g. cameras, microphones and accelerometers), images, sounds and movements, combined with state-of-the-art algorithms, can be used to automatically monitor animal behavior or other animal-based measures. New technological innovations are thus appearing, although their practical implementation for welfare assessment is still limited. PLF may be well suited to assess the welfare of individual animals. A number of publications have pointed out the potential use of PLF for welfare monitoring (Nadimi et al., 2012; Rutter, 2014; Shalloo et al., 2018; Werner et al., 2019).

In this chapter an overview will be provided on current knowledge and future potential of PLF technology to enhance dairy cattle welfare. The potential of PLF technology will be described linked to the Five Domains framework (Mellor, 2017). Therefore, first this framework will be shortly described, after which the potential of the use of PLF for each of the domains will be given. Finally, the need for data processing algorithms will be described and discussed.

## 2 The five domains model for welfare assessment

The Five Freedoms framework of the UK Farm Animal Welfare Council (FAWC) is one of the most widely used frameworks used in relation to animal welfare (Mellor, 2017). Since animal welfare contains many aspects, attempts to create a simple definition of good animal welfare typically share two main components: that animals are physically healthy and that they are largely free from mental suffering. However, these simplistic definitions hide the fact that a multitude of factors contribute to animal welfare, and this is arguably the main reason it is not possible to use one common, simple definition. In contrast, the Five Freedoms framework reflects the multitude of factors that impact animal welfare.

With the Five Freedoms framework, the strengths and weaknesses of animal husbandry systems can be judged (Webster, 2008). The Five Freedoms framework has and is extensively used as the basis for animal welfare programs (e.g. McCulloch, 2013). In order to address limitations of the Five Freedoms framework, but staying closely connected, the Five Domains framework was developed (Mellor, 2015, 2017). Fogarty et al. (2019) have used this framework to evaluate the potential of PLF to monitor the welfare of sheep.

The Five Domains framework (Table 1) consists of five areas, domains, in which welfare can be compromised or enhanced. Three domains – nutrition, environment and health – contain survival-related factors. One domain contains factors regarding the external situation and one domain contains an overarching assessment of the influence of domains 1–4 on the affective experience of the animal. Because of the structure of this framework, animal welfare features in the fifth domain (mental state) are linked to the other four domains. For instance, presence of disease or injury (features in the health domain) is associated with pain, debility and sickness, which are features in the fifth domain (mental state). Because of that association, in the following four paragraphs, the potential application of PLF systems to monitor features in these four domains (nutrition, environment, health and behavior) will be described including aspects linked to the fifth domain (mental state).

## 3 The nutrition domain

One of the first applications of PLF was the introduction of automated concentrate feeders. The development started already in the 1960s (Harshbarger et al., 1965), but started to become widely applied in the 1980s (Van Asseldonk et al., 1998). In confinement systems, the use of automated feeding ensures intake of sufficient feed and that way prevents hunger. Although the currently used water systems cannot be seen as PLF, they are developed and designed in such a way that water is always available, as long as there are enough water troughs. One aspect in which sensor systems may play a role is to ensure a proper functioning of the water and feed systems. Water flow systems are used in pig farming and are even used in PLF applications (e.g. Jensen et al., 2017).

A specific form of feeding automation is the use of automated calf feeders (Jorgensen et al., 2017; Medrano-Galarza et al., 2018). Combined with a weighing scale, this allows for decision support regarding calf feeding and calf rearing decisions (Cantor et al., 2020; Eslamizad et al., 2018). Typically of these automated feeding and drinking systems is that they were developed to either improve production efficiency or reduce the use of labor (automation). However, by performing those tasks, these systems also support animal welfare by ensuring a freedom of hunger and thirst. Moreover, when the decision algorithms behind these systems would take balance of diet into account, these

**Table 1** Key features and examples of the extended five domains model for welfare assessment (Mellor, 2015) as used by (Fogarty et al., 2019)

**Physical/functional domains**

| | 1 Nutrition | 2 Environment | 3 Health | 4 Behavior |
|---|---|---|---|---|
| | Survival-related factors | | | Situation-related factors |
| Restrictions: | Water or food intake | | | |
| | Food quality and variety | | | |
| | Voluntary overeating | | | |
| Unavoidable/imposed conditions: | | Thermal extremes | | |
| | | Close confinement | | |
| | | Unpredictable events | | |
| Presence of: | | | Disease | |
| | | | Injury | |
| | | | Functional impairment | |
| Exercise of 'agency' impeded by: | | | | Choices markedly restricted |
| | | | | Constraints on anima-animal interaction |
| Opportunities: | Drink enough water | | | |
| | Eat a balanced/varied diet | | | |
| Available conditions: | | Thermally tolerable | | |
| | | Space for free movement | | |
| | | Normal environment variation | | |
| Presence of: | | | Appropriate body condition score | |
| | | | Good fitness level | |
| 'Agency' exercised by: | | | | Available engaging choices |
| | | | | Bonding |
| | | | | Play |

**Affective experience domain**

**5 Mental state**

| | 1 Nutrition | 2 Environment | 3 Health | 4 Behavior |
|---|---|---|---|---|
| Negative: | Thirst | | | |
| | Hunger | | | |
| Negative forms of discomfort: | | Thermal | | |
| | | Respiratory | | |
| | | Auditory | | |
| Negative: | | | Pain | |
| | | | Debility | |
| | | | Sickness | |
| Negative: | | | | Anger |
| | | | | Boredom |
| | | | | Helplessness |
| Positive: | Quenching thirst | | | |
| | Pleasure of different tastes/smells | | | |
| Positive forms of comfort: | | Thermal | | |
| | | Respiratory | | |
| | | Auditory | | |
| Positive: | | | Comforts of good health and high functional capacity | |
| Positive calmness: | | | | Engaged |
| | | | | Maternally rewarded |
| | | | | Playfulness |

aspects of dairy cattle welfare may also be supported by systems that are in use anyway. For instance, by monitoring rumination and activity, potential negative effects of dietary changes may be detected (Giaretta et al., 2019).

One aspect about dairy cattle nutrition requires specific attention. That is the issue of roughage intake. In confinement systems, the total roughage intake for a group of cows is often known. However, individual intake of roughage is lacking and individual cows vary in their actual roughage intake. Individual cow roughage intake systems are available (Devir et al., 1996), but these are only used in experimental settings due to the complexity of the system and the price. Recently, using 3D accelerometers, it is shown that it is possible to get insight in roughage intake of dairy cattle (Carpinelli et al., 2019; Zambelis et al., 2019; Reiter et al., 2018) as well as the feeding (Byskov et al., 2015; Ruuska et al., 2016) and drinking behavior of cattle (Williams et al., 2019). Positioning sensor systems may add to the prediction of feeding behavior (Barker et al., 2018; Mattachini et al., 2016). Also in grazing systems, roughage intake has been studied with sensor systems (Jemila and Priyadharsini, 2018; Oudshoorn et al., 2013). Using neck-mounted 3D accelerometers, it is possible to estimate grass intake (Andriamandroso et al., 2017). These developments open possibilities to monitor the feed and water intake and ensure the freedom of hunger and thirst from the cow behavior's point of view.

## 4 The environment domain

Animal welfare aspects in the environment domain are often related to the design and layout of a barn and PLF systems do not play a large role in that. A proper barn design may provide enough space to cows to guarantee free movement. Or, in climates with a risk of heat stress, barn designs do have equipment to mitigate the effects of the high temperature and humidity. By keeping cows in well-designed barns, animal comfort may be guaranteed as good as possible. Moreover, by having measuring equipment (sensors) in the barn to monitor the barn climate, a farmer can monitor that climate and take action when needed.

However, more and more it is important to evaluate the actual well-being of cows rather than the conditions. For instance, climatic circumstances may lead to heat stress in dairy cattle. Recent work has been carried out to monitor the thermal status of dairy cattle using 3D accelerometers (Bar et al., 2019). Algorithms were developed to monitor the breathing of cows and monitoring 'heavy breathing', an indication of heat stress. Based on this indicator, cooling interventions could be undertaken.

The space for free movement is another important aspect in the environment domain. By measuring the movements of cows, it will be possible to monitor whether cows can move freely. Especially accelerometers are able to carry out that task (Martiskainen et al., 2009). Currently a number of PLF technologies are available for this type of monitoring (e.g. Borchers et al., 2016). The freedom of movement is also associated with the ease of walking. The construction and the cleanliness of the floor is an important aspect for ease of walking. Also, suboptimal mobility leads to impairment of the ease of walking. Suboptimal mobility is associated with claw problems (O'Connor et al., 2019) and does also affect the ease of walking. Suboptimal mobility can be monitored by the use of different types of PLF sensor systems (e.g. Barker et al., 2018; Beer et al., 2016; Nechanitzky et al., 2016; Pastell et al., 2008, 2009; Thorup et al., 2015; Van de Gucht et al., 2017). It seems that the performance of these systems, in terms of sensitivity and specificity is not yet good enough to manage individual cows, but most probably these types of systems can be used to monitor the ease of movement of the whole herd.

## 5 The health domain

By ensuring good dairy cattle health, we can safeguard them from pain, debility and sickness. Animal health is, therefore, an important component of animal welfare. The most important diseases with regard to animal welfare are the so-called production diseases: diseases that are associated with dairy production and which are often relatively highly prevalent. The most important of these diseases are mastitis, claw disorders and metabolic disorders.

Until now, PLF applications have especially been studied to monitor dairy cattle health (Rutten et al., 2013). The most studied disease is mastitis. Obviously because of the (economic) importance of mastitis (Hogeveen et al., 2019) as well as because of the introduction of automatic milking, where mastitis detection could not anymore be done by a milker. Therefore, sensor systems were needed to replace the visual observations of the milker (Hogeveen et al., 2010). Mostly these sensors utilized electrical conductivity but recently somatic cell count sensors are becoming available as well (Dalen et al., 2019; Deng et al., 2020).

Foot disorders, associated with suboptimal mobility, are costly (Bruijnis et al., 2010) and might be more prevalent than mastitis; the prevalence of suboptimal mobility was more than 35% in a recent study (O'Connor et al., 2019). Since foot disorders are associated with pain it is an important disease from a welfare point of view (Bruijnis et al., 2012). Although some work has been carried out on PLF systems to detect suboptimal mobility (see the previous paragraph), it seems that the development of PLF applications for suboptimal mobility does not have a priority. Potentially that is because that work is aimed at the detection of cows with suboptimal mobility, a task that farmers can also carry out visually. No work has been carried out so far on using PLF to monitor suboptimal mobility at the herd level, which could be an interesting welfare monitoring application.

Some work has been carried out on the development of PLF systems to detect cows with metabolic disorders, both subclinical and clinical. A number of predictors are available (LeBlanc et al., 2005; Overton et al., 2017). One of the potential predictors is rumination, a predictor that is used a number of times to detect metabolic disorders using PLF (Kroger et al., 2016; Molfino et al., 2017; Reiter et al., 2018; Schirmann et al., 2009), potentially in combination with activity monitoring (Stangaferro et al., 2016). Often this work is done using neck or ear-mounted 3D accelerometers but rumen bolus sensors (Benaissa et al., 2019; Hamilton et al., 2019; Sturm et al., 2020), in-line beta-hydroxybutyrate sensors (Nielsen et al., 2005), electronic nose (Mottram, 1997; Mottram et al., 1999), the milk fat-to-protein ratio (King et al., 2019), lying (Kaufman et al., 2016) or behavioral patterns (Wagner et al., 2020) are also used.

Interestingly, all PLF applications in the health domain that are described so far are aimed at detecting a suboptimal situation (disease) at the cow level. Detection of a suboptimal situation may be used to initiate an intervention at cow-level, for example, treatment. The uptake of these systems is not always high (Steeneveld and Hogeveen, 2015; Borchers and Bewley, 2015; Stone, 2020) which may be associated by either a lack of clear intervention options or by a lack of accuracy of the PLF system in comparison with the farmers' accuracy. Until now, no authors have been looking at the value of PLF systems in monitoring health at the herd level. This type of monitoring is also useful with regard to the monitoring of animal welfare. By monitoring disease occurrence over time, a farmer and other interested stakeholders have a real-time overview of the health status of cows on an individual farm.

The usefulness of PLF has also been studied regarding bovine respiratory disease. Most of the studies were aimed at respiratory disease in youngstock, using 3D accelerometers (Swartz et al., 2017) and sound recordings to measure coughing (Carpentier et al., 2018; Ferrari et al., 2010; Vandermeulen et al., 2016).

Besides monitoring the presence of the diseases, the negative aspect of the health domain, PLF systems may also be helpful in monitoring positive aspects such as the body condition score (e.g. Bercovich et al., 2013; Mullins et al., 2019), milk production level as expected (e.g. Adriaens et al., 2018; Jensen et al., 2016), absence of fever (Kou et al., 2017), feeding behavior (Gonzalez et al., 2008; Knauer et al., 2018; Svensson and Jensen, 2007) or the absence of any disease signs (Eckelkamp and Bewley, 2020; King and DeVries, 2018). It may even be possible to use routinely available milk shipment data to monitor disease at the herd level (Fall et al., 2018).

## 6 The behavior domain

As indicated earlier, sensor systems so-far have been developed and studied with regard to animal, disease and production management. For dairy farmers, there is a (financial) value in those aspects of animal management, while animal behavior, as such, does not contain financial value. However, animal behavior is an essential aspect of animal welfare. Animals should be able to perform their natural behavior to prevent anger, boredom and helplessness. Besides the often mentioned feeding, standing and lying behavior, which are essential behavioral aspects, also social behavior (interaction with other cows) is an important aspect in the behavior domain.

PLF systems may be very suitable to monitor animal behavior (Rushen et al., 2012). Especially 3D accelerometers are able to distinguish between several types of behavior (standing, eating, lying, walking). Since lying is an important welfare indicator and can also be used to evaluate the comfort of stalls within barns, a number of studies have focused on algorithms to evaluate lying time of cows (e.g. Kok et al., 2015; Robert et al., 2011). Some other studies have more generally studied the use of PLF systems to explore the time budgets of dairy cattle (e.g. Arcidiacono et al., 2017; Benaissa et al., 2019; Grinter et al., 2019; Jaeger et al., 2019; Tamura et al., 2019).

Adding cow location sensors to the accelerometer sensors does improve the precision of the cow behavior monitoring (Wang et al., 2019). No studies have been carried out on monitoring cattle interaction (except for work on estrus detection), but systems that do monitor the indoor position of cows can be an interesting tool to monitor social interaction between cows (Pastell et al., 2018). Another potential PLF system that may be used to monitor social interaction between animals is the use of systems such as underdevelopment for the Coronavirus contact-tracing apps working with mobile phone technology (Fraser et al., 2020). Such systems allow the evaluation of the number of cows one cow has been interacting with, potentially in combination with the place where that interaction happened. An example of utilizing the potential of monitoring the interaction between cows was given by Huzzey et al. (2014), who looked at social competition in front of automated feeding systems.

Grazing behavior is another aspect that pleases a role. PLF may be useful to enable efficient grazing systems by virtual fencing (Lomax et al., 2019). Just as with in-house feeding behavior, 3D accelerometers are also used to evaluate grazing behavior (Werner et al., 2019; Pereira et al., 2020; Gonzalez et al., 2015; Poulopoulou et al., 2019). Especially with this aspect, cow position information can be of additional value to monitor where cows have been grazing besides how cows have been grazing. From a regulatory perspective the monitoring of grazing may also be interesting since in some countries or with some dairy-processing companies, farmers are rewarded when their cows graze, which means that a system should be in place to assess grazing of cows.

# 7 The need for algorithms to monitor dairy cattle welfare

Since much of the research done so far was not aimed specifically at monitoring animal welfare, we chose to give an overview of the potential use of PLF in monitoring welfare and to not try to provide the accuracy with which monitoring dairy cattle welfare can be carried out. Nor did we try to be complete in all references or all PLF systems. Given the currently described PLF systems, a summary of the PLF systems that may play a role can be created:

- Accelerometers. These may be placed in the ear, around the leg or the neck. Their basic application is estrus detection and they are widely used in dairy farms. Depending on their position they can be used in monitor feeding and other behavior of dairy cattle and be useful to evaluate a wide range of features relating to animal welfare.
- Cow-positioning electronics. These may be placed in the ear, around the leg or in the neck, often in combination with accelerometers. Cow-positioning sensors are in practice and are used to find specific cows, but can be useful to evaluate social interaction between cows and some types of behavior.
- Temperature sensors. These may be placed in the ear, in the milk line and in the rumen. Temperature measurements can be used in monitoring animal health and heat stress.
- pH sensors. This sensor may be placed in the rumen and can be used to monitor metabolic status of cows, aimed at detecting subclinical acidosis, or deviations in the feeding ration.
- Cameras. A number of different cameras may be utilized in monitoring animal welfare. In an overview placement, cameras can be used to monitor cows' positions as well as social interaction and discomfort. Closer to the cow they can be used to evaluate body condition score and thus the general well-being of a cow. In another position, cameras may be used to monitor mobility of cows. Finally, thermal cameras may be used to monitor cows' health.
- Sensors measuring parameters in milk. A wide range of sensors are available, mostly aimed at monitoring udder health and milk production (milk yield, fat and protein percentage). The production parameters can be used to monitor the absence of disease.

Although all of the above sensor systems have been described in literature and have shown to be potentially useful for a specific aspect of dairy farm management, specific algorithms should be developed to enable the use of these sensors in animal welfare monitoring. The developments in machine learning and big data analytics have been extensive in the past years (Morota et al., 2018). It is expected that the machine-learning toolbox is sufficiently filled to create proper algorithms to monitor dairy cattle welfare based upon PLF systems. However, before useful algorithms can be developed, the performance needs for those algorithms should be known. Until now, most machine-learning algorithms are aimed at the detection of cases of disease in individual animals. Depending on the type of disease, the prevalence of disease, the potential welfare and production consequences of missing a disease and the costs for intervention, specifications for the detection performance may be defined. Test characteristics typically consist of the sensitivity, the specificity and the time window in which the disease should be detected. However, until now, minimal test characteristics have only been provided for one disease: clinical mastitis (Hogeveen et al., 2010).

For animal welfare issues, interventions are most probably provided at the herd level, rather than at the individual cow level. Therefore, the individual cow assessment of animal welfare features need to be aggregated. Performance of algorithms to monitor dairy cattle welfare at the herd level should be evaluated at the herd level and not at the individual cow level. Imperfections at the cow level algorithms to assess welfare, if random, may be less important when individual cow assessments are aggregated at the herd level.

Besides the use of PLF for production monitoring and optimization and for animal disease management, PLF may also be used to assess animal welfare on dairy farms. Such an assessment may be used by the farm management to intervene. Farm managers should receive an alert when dairy cattle behavior, for instance, lying behavior is changing. These alerts could be linked to key performance indicators (KPIs) related to animal welfare. Such KPIs can consist of the proportion of cows with abnormal behavior, or the proportion of cows potentially suffering from a negative situation, such as heat stress, pain, hunger or thirst (aspects from the fifth domain, the mental state). Linked to these KPIs, algorithms should be provided that provide alerts. The needed performance for such alerts depends on the severity of the feature. Issues with hunger/thirst or severe pain should be alerted very quickly and should have a high sensitivity. Issues regarding changes in cow behavior may last longer. Issues where quick intervention is possible may need different performance than issues that need further diagnostics. Typically algorithms that may be used for such tasks could be based on finding a deviation in the trend, such as dynamic linear programming (e.g. Cornou et al., 2011, 2014) or statistical process control (e.g. Huybrechts et al., 2014). But before algorithms can be trained, it would be good that experts in the fields of animal welfare, farm management, animal health and data management would get together to work out the demands for algorithms given the available PLF systems. The scientific knowledge so far is not sufficient in that respect. As an example, for mastitis management a number of experts have been together and worked out demands for PLF systems to management of mastitis at various levels, beyond the detection of clinical mastitis (Hogeveen et al., 2020).

# 8 Conclusion

At the moment, the availability of more and more, relatively cheap technology in combination with the processing power of modern computers is bringing a large change to the management of dairy farms (Cabrera et al., 2020; Barkema et al., 2015). Precision livestock farming has brought a number of sensor and automated data acquisition technologies to commercial farms (Berckmans, 2014). Most of these sensor and data acquisition systems are primarily developed to improve production efficiency. As long as precision livestock technology is not associated with premium prices, financial efficiency of these systems needs to be achieved through improvements in resource-use efficiency. However, many of the data acquired by precision technology are pertinent to welfare assessment (Rutter, 2014) and might be a basis for premium prices. Moreover, by linking the precision technology to adequate animal-based decision support systems, optimal use of (environmental) resources may be supported (Shalloo et al., 2018), improving the sustainability of livestock production.

There is an increasing societal demand for animal welfare assessment (Winckler, 2019). Over time, processors and/or retailers may demand that all livestock products come with objective, sensor-derived farm data to demonstrate they were produced to acceptable welfare standards. Current resource-based assessment systems do have a value in enabling a systematic evaluation of animal welfare. However, it has been criticized for a number of reasons: (1) using too few animal-based measures, (2) time-consuming yet infrequent assessments and (3) inflexible weighting of different welfare factors (McCulloch, 2013). Besides the inflexibility of the current weighting of the criteria, they do not necessarily conform to the public's perceptions of the relative importance of the factors that contribute to animal welfare. For example, consumers seem to accept quite a number of negative animal welfare aspects (e.g. disease) as long as the animals are free ranged (Bergstra et al., 2017). In other words, animal welfare assessment systems should be based on a flexible system of weighting the various aspects of animal welfare, in order to meet consumers' preferences as well as scientific knowledge. Moreover, in order to be useful, assessment of animal welfare should be animal-based and easy to perform in order to be used. PLF systems may, as is shown in this chapter, be of good help in obtaining and monitoring an overall view of dairy cattle welfare.

Currently, the uptake of precision technologies available varies between different livestock systems and European regions. Although the greatest uptake of PLF technologies has, to date, been in intensively managed livestock production systems with much attention for individual animals (e.g. more than 15% of Dutch dairy farmers have 3D accelerometers, Steeneveld and Hogeveen, 2015), there is no fundamental reason why they could not also be used in more extensively managed systems (e.g. Gonzalez et al., 2015; Rutter, 2014). The benefit of PLF systems to monitor animal welfare is only partly for a dairy farmer. That part is linked to the use of PLF systems to make production more efficient and to reduce the negative consequences of animal disease. Although farmers have an intrinsic interest in animal welfare (Valeeva et al., 2007), they will most likely not invest much in sensor systems to monitor animal welfare. By utilizing PLF systems that are used by farmers for other purposes (e.g. 3D accelerometers which are very cost-efficient to be used for estrus detection (Rutten et al., 2013), the animal welfare monitoring can be carried out without any additional investments in hardware. Typically other stakeholders in the dairy value chain may be interested in utilizing such PLF-based welfare monitoring systems, for instance, for milk quality certification programs. Business models to pay for the infrastructure and the additional software should be developed. Finally, we should also note that, from an ethical point of view there are some remarks on the use of PLF systems, because they 'objectify' the animal (Bos et al., 2018).

In conclusion, based on the Five Domains framework, an overview has been given on the potential use of PLF systems to monitor animal welfare. Currently, sensor technology is available but has hardly been used to monitor animal welfare. We are at the start of these developments. Much is still needed in defining the needs for animal welfare monitoring. That work should provide the foundation on which algorithms can be created that utilize the PLF technology data for monitoring animal welfare.

## 9 Where to look for further information

In this chapter, we provided an overview of the precision livestock techniques that may be used to monitor animal welfare. We showed that with the currently available techniques it is possible to monitor animal welfare. Yet, the scientific literature on this topic is highly fragmented, the association between single animal welfare aspects and data collected by sensor systems is provided, often from the point of view of animal disease management (disease detection) or measuring production efficiency (looking at feed intake and feeding behavior). Moreover, studies are either carried out from a more technological point of view (for example aimed at improved sensing technology or algorithms) or from a more animal welfare point of view. A first step for future research would be the development of a concise animal welfare monitoring framework from the point of view of precision livestock technology. Such a framework should be developed by a trans-disciplinary group of researchers representing a number of research and implementation disciplines. Within such a framework researchers, and within their own discipline may work out elements. The information brought together in this chapter can be used as a starting point.

# 10 References

Adriaens, I., Huybrechts, T., Aernouts, B., Geerinckx, K., Piepers, S., De Ketelaere, B. and Saeys, W. 2018. Method for short-term prediction of milk yield at the quarter level to improve udder health monitoring. *Journal of Dairy Science* 101(11), 10327-10336.

Andriamandroso, A. L. H., Lebeau, F., Beckers, Y., Froidmont, E., Dufrasne, I., Heinesch, B., Dumortier, P., Blanchy, G., Blaise, Y. and Bindelle, J. 2017. Development of an open-source algorithm based on inertial measurement units (IMU) of a smartphone to detect cattle grass intake and ruminating behaviors. *Computers and Electronics in Agriculture* 139, 126-137.

Arcidiacono, C., Porto, S. M. C., Mancino, M. and Cascone, G. 2017. Development of a threshold-based classifier for real-time recognition of cow feeding and standing behavioural activities from accelerometer data. *Computers and Electronics in Agriculture* 134, 124-134.

Bar, D., Kaim, M., Flamenbaum, I., Hanochi, B. and Toaff-Rosenstein, R. L. 2019. Technical note: accelerometer-based recording of heavy breathing in lactating and dry cows as an automated measure of heat load. *Journal of Dairy Science* 102(4), 3480-3486.

Barkema, H. W., Von Keyserlingk, M. A. G., Kastelic, J. P., Lam, T. J., Luby, C., Roy, J. P., LeBlanc, S. J., Keefe, G. P. and Kelton, D. F. 2015. Invited review: changes in the dairy industry affecting dairy cattle health and welfare. *Journal of Dairy Science* 98(11), 7426-7445.

Barker, Z. E., Diosdado, J. A. V., Codling, E. A., Bell, N. J., Hodges, H. R., Croft, D. P. and Amory, J. R. 2018. Use of novel sensors combining local positioning and acceleration to measure feeding behavior differences associated with lameness in dairy cattle. *Journal of Dairy Science* 101(7), 6310-6321.

Beer, G., Alsaaod, M., Starke, A., Schuepbach-Regula, G., Muller, H., Kohler, P. and Steiner, A. 2016. Use of extended characteristics of locomotion and feeding behavior for automated identification of lame dairy cows. *PLoS ONE* 11(5), e0155796.

Benaissa, S., Tuyttens, F. A. M., Plets, D., Cattrysse, H., Martens, L., Vandaele, L., Joseph, W. and Sonck, B. 2019. Classification of ingestive-related cow behaviours using RumiWatch halter and neck-mounted accelerometers. *Applied Animal Behaviour Science* 211, 9-16.

Berckmans, D. 2014. Precision livestock farming technologies for welfare management in intensive livestock systems. *Revue Scientifique et Technique* 33(1), 189-196.

Bercovich, A., Edan, Y., Alchanatis, V., Moallem, U., Parmet, Y., Honig, H., Maltz, E., Antler, A. and Halachmi, I. 2013. Development of an automatic cow body condition scoring using body shape signature and fourier descriptors. *Journal of Dairy Science* 96(12), 8047-8059.

Bergstra, T., Hogeveen, H., Kuiper, W. E., Oude Lansink, A. G. J. M. and Stassen, E. N. 2017. Attitudes of Dutch citizens toward sow husbandry with regard to animals, humans, and the environment. *Anthrozoös* 30(2), 195-211.

Borchers, M. R. and Bewley, J. M. 2015. An assessment of producer precision dairy farming technology use, prepurchase considerations, and usefulness. *Journal of Dairy Science* 98(6), 4198-4205.

Borchers, M. R., Chang, Y. M., Tsai, I. C., Wadsworth, B. A. and Bewley, J. M. 2016. A validation of technologies monitoring dairy cow feeding, ruminating, and lying behaviors. *Journal of Dairy Science* 99(9), 7458-7466.

Bos, J. M., Bovenkerk, B., Feindt, P. H. and Van Dam, Y. K. 2018. The quantified animal: precision livestock farming and the ethical implications of objectification. *Food Ethics* 2(1), 77-92.

Botreau, R., Bonde, M., Butterworth, A., Perny, P., Bracke, M. B. M., Capdeville, J. and Veissier, I. 2007. Aggregation of measures to produce an overall assessment of animal welfare. Part 1: a review of existing methods. *Animal: An International Journal of Animal Bioscience* 1(8), 1179-1187.

Bruijnis, M. R. N., Beerda, B., Hogeveen, H. and Stassen, E. N. 2012. Assessing the welfare impact of foot disorders in dairy cattle by a modeling approach. *Animal: An International Journal of Animal Bioscience* 6(6), 962-970.

Bruijnis, M. R. N., Hogeveen, H. and Stassen, E. N. 2010. Assessing economic consequences of foot disorders in dairy cattle using a dynamic stochastic simulation model. *Journal of Dairy Science* 93(6), 2419-2432.

Byskov, M. V., Nadeau, E., Johansson, B. E. O. and Norgaard, P. 2015. Variations in automatically recorded rumination time as explained by variations in intake of dietary fractions and milk production, and between-cow variation. *Journal of Dairy Science* 98(6), 3926-3937.

Cabrera, V. E., Barrientos-Blanco, J. A., Delgado, H. and Fadul-Pacheco, L. 2020. Symposium review: real-time continuous decision making using big data on dairy farms. *Journal of Dairy Science* 103(4), 3856-3866.

Cantor, M. C., Pertuisel, C. H. and Costa, J. H. C. 2020. Technical note: estimating body weight of dairy calves with a partial-weight scale attached to an automated milk feeder. *Journal of Dairy Science* 103(2), 1914-1919.

Carpentier, L., Berckmans, D., Youssef, A., Berckmans, D., Van Waterschoot, T., Johnston, D., Ferguson, N., Earley, B., Fontana, I., Tullo, E., Guarino, M., Vranken, E. and Norton, T. 2018. Automatic cough detection for bovine respiratory disease in a calf house. *Biosystems Engineering* 173, 45-56.

Carpinelli, N. A., Rosa, F., Grazziotin, R. C. B. and Osorio, J. S. 2019. Technical note: a novel approach to estimate dry matter intake of lactating dairy cows through multiple on-cow accelerometers. *Journal of Dairy Science* 102(12), 11483-11490.

Cornou, C., Lundbye-Christensen, S. and Kristensen, A. R. 2011. Modelling and monitoring sows' activity types in farrowing house using acceleration data. *Computers and Electronics in Agriculture* 76(2), 316-324.

Cornou, C., Ostergaard, S., Ancker, M. L., Nielsen, J. and Kristensen, A. R. 2014. Dynamic monitoring of reproduction records for dairy cattle. *Computers and Electronics in Agriculture* 109, 191-194.

Dalen, G., Rachah, A., Nørstebø, H., Schukken, Y. H. and Reksen, O. 2019. The detection of intramammary infections using online somatic cell counts. *Journal of Dairy Science* 102(6), 5419-5429.

Deng, Z. J., Hogeveen, H., Lam, T. J. G. M., Van Der Tol, R. and Koop, G. 2020. Performance of online somatic cell count estimation in automatic milking systems. *Frontiers in Veterinary Science* 7, 221.

Devir, S., Hogeveen, H., Hogewerf, P. H., Ipema, A. H., Ketelaardelauwere, C. C. K., Rossing, W., Smits, A. C. and Stefanowska, J. 1996. Design and implementation of a system for automatic milking and feeding. *Canadian Agricultural Engineering* 38, 107-113.

Eckelkamp, E. A. and Bewley, J. M. 2020. On-farm use of disease alerts generated by precision dairy technology. *Journal of Dairy Science* 103(2), 1566-1582.

Eslamizad, M., Tummler, L. M., Derno, M., Hoch, M. and Kuhla, B. 2018. Technical note: development of a pressure sensor-based system for measuring rumination time in pre-weaned dairy calves. *Journal of Animal Science* 96(11), 4483-4489.

Fall, N., Ohlson, A., Emanuelson, U. and Dohoo, I. 2018. Exploring milk shipment data for their potential for disease monitoring and for assessing resilience in dairy farms. *Preventive Veterinary Medicine* 154, 23-28.

Ferrari, S., Piccinini, R., Silva, M., Exadaktylos, V., Berckmans, D. and Guarino, M. 2010. Cough sound description in relation to respiratory diseases in dairy calves. *Preventive Veterinary Medicine* 96(3-4), 276-280.

Fogarty, E. S., Swain, D. L., Cronin, G. M. and Trotter, M. 2019. A systematic review of the potential uses of on-animal sensors to monitor the welfare of sheep evaluated using the five domains model as a framework. *Animal Welfare* 28(4), 407-420.

Fraser, C., Abeler-Doerner, L., Ferretti, L., Parker, M., Kendall, M. and Bonsall, D. 2020. *Digital Contact Tracing: Comparing the Capabilities of Centralised and Decentralised Data Architectures to Effectively Suppress the COVID-19 Epidemic Whilst Maximising Freedom of Movement and Maintaining Privacy*, Nature Publishing, preprint https:// github.com/BDI-pathogens/covid-19_instant_tracing/blob/master/Centralised%20 and%20decentralised%20systems%20for%20contact%20tracing.pdf.

Giaretta, E., Mordenti, A. L., Canestrari, G., Palmonari, A. and Formigoni, A. 2019. Automatically monitoring of dietary effects on rumination and activity of finishing heifers. *Animal Production Science* 59(10), 1931-1940.

Gonzalez, L. A., Bishop-Hurley, G. J., Handcock, R. N. and Crossman, C. 2015. Behavioral classification of data from collars containing motion sensors in grazing cattle. *Computers and Electronics in Agriculture* 110, 91-102.

Gonzalez, L. A., Tolkamp, B. J., Coffey, M. P., Ferret, A. and Kyriazakis, I. 2008. Changes in feeding behavior as possible indicators for the automatic monitoring of health disorders in dairy cows. *Journal of Dairy Science* 91(3), 1017-1028.

Grinter, L. N., Campler, M. R. and Costa, J. H. C. 2019. Technical note: validation of a behavior-monitoring collar's precision and accuracy to measure rumination, feeding, and resting time of lactating dairy cows. *Journal of Dairy Science* 102(4), 3487-3494.

Halachmi, I., Guarino, M., Bewley, J. and Pastell, M. 2019. Smart animal agriculture: application of real-time sensors to improve animal well-being and production. *Annual Review of Animal Biosciences* 7, 403-425.

Hamilton, A. W., Davison, C., Tachtatzis, C., Andonovic, I., Michie, C., Ferguson, H. J., Somerville, L. and Jonsson, N. N. 2019. Identification of the rumination in cattle using support vector machines with motion-sensitive bolus sensors. *Sensors* 19(5), 1165.

Harshbarger, K. E., Olver, E. F. and Shove, G. C. 1965. Effects of an automatic water-concentrate feeder on milk production. *Journal of Dairy Science* 48, 794.

Hogeveen, H., Kamphuis, C., Steeneveld, W. and Mollenhorst, H. 2010. Sensors and clinical mastitis-the Quest for the perfect alert. *Sensors* 10(9), 7991-8009.

Hogeveen, H., Klaas, I. C., Dalen, G., Honig, H., Zecconi, A., Kelton, D. F. and Sanchez Mainar, M. 2020. Novel approaches to manage mastitis support by sensor systems. Submitted for publication.

Hogeveen, H., Steeneveld, W. and Wolf, C. A. 2019. Production diseases reduce the efficiency of dairy production: a review of the results, methods, and approaches

Regarding the economics of mastitis. *Annual Review of Resource Economics* 11(1), 289-312.

Huybrechts, T., Mertens, K., De Baerdemaeker, J., De Ketelaere, B. and Saeys, W. 2014. Early warnings from automatic milk yield monitoring with online synergistic control. *Journal of Dairy Science* 97(6), 3371-3381.

Huzzey, J. M., Weary, D. M., Tiau, B. Y. F. and Von Keyserlingk, M. A. G. 2014. Short communication: automatic detection of social competition using an electronic feeding system. *Journal of Dairy Science* 97(5), 2953-2958.

Jaeger, M., Brugemann, K., Brandt, H. and Konig, S. 2019. Associations between precision sensor data with productivity, health and welfare indicator traits in native black and white dual-purpose cattle under grazing conditions. *Applied Animal Behaviour Science* 212, 9-18.

Jemila, J. S. and Priyadharsini, S. S. 2018. A sensor-based forage monitoring of grazing cattle in dairy farming. *International Journal on Smart Sensing and Intelligent Systems* 11(1), 1-9.

Jensen, D. B., Hogeveen, H. and De Vries, A. 2016. Bayesian integration of sensor information and a multivariate dynamic linear model for prediction of dairy cow mastitis. *Journal of Dairy Science* 99(9), 7344-7361.

Jensen, D. B., Toft, N. and Kristensen, A. R. 2017. A multivariate dynamic linear model for early warnings of diarrhea and pen fouling in slaughter pigs. *Computers and Electronics in Agriculture* 135, 51-62.

Jorgensen, M. W., Adams-Progar, A., De Passille, A. M., Rushen, J., Godden, S. M., Chester-Jones, H. and Endres, M. I. 2017. Factors associated with dairy calf health in automated feeding systems in the Upper Midwest United States. *Journal of Dairy Science* 100(7), 5675-5686.

Kaufman, E. I., LeBlanc, S. J., Mcbride, B. W., Duffield, T. F. and Devries, T. J. 2016. Short communication: association of lying behavior and subclinical ketosis in transition dairy cows. *Journal of Dairy Science* 99(9), 7473-7480.

King, M. T. M. and DeVries, T. J. 2018. Graduate Student Literature Review: detecting health disorders using data from automatic milking systems and associated technologies. *Journal of Dairy Science* 101(9), 8605-8614.

King, M. T. M., Duffield, T. F. and Devries, T. J. 2019. Short communication: assessing the accuracy of inline milk fat-to-protein ratio data as an indicator of hyperketonemia in dairy cows in herds with automated milking systems. *Journal of Dairy Science* 102(9), 8417-8422.

Knauer, W. A., Godden, S. M., Dietrich, A., Hawkins, D. M. and James, R. E. 2018. Evaluation of applying statistical process control techniques to daily average feeding behaviors to detect disease in automatically fed group-housed preweaned dairy calves. *Journal of Dairy Science* 101(9), 8135-8145.

Kok, A., Van Knegsel, A. T. M., Van Middelaar, C. E., Hogeveen, H., Kemp, B. and De Boer, I. J. M. 2015. Technical note: validation of sensor-recorded lying bouts in lactating dairy cows using a 2-sensor approach. *Journal of Dairy Science* 98(11), 7911-7916.

Kou, H. X., Zhao, Y. Q., Ren, K., Chen, X. L., Lu, Y. Q. and Wang, D. 2017. Automated measurement of cattle surface temperature and its correlation with rectal temperature. *PLoS ONE* 12(4), e0175377.

Kroger, I., Humer, E., Neubauer, V., Kraft, N., Ertl, P. and Zebeli, Q. 2016. Validation of a noseband sensor system for monitoring ruminating activity in cows under different feeding regimens. *Livestock Science* 193, 118-122.

LeBlanc, S. J., Leslie, K. E. and Duffield, T. F. 2005. Metabolic predictors of displaced abomasum in dairy cattle. *Journal of Dairy Science* 88(1), 159-170.

Lomax, S., Colusso, P. and Clark, C. E. F. 2019. Does virtual fencing work for grazing dairy cattle? *Animals: An Open Access Journal from MDPI* 9(7), 429.

Martiskainen, P., Jarvinen, M., Skon, J. P., Tiirikainen, J., Kolehmainen, M. and Mononen, J. 2009. Cow behaviour pattern recognition using a three-dimensional accelerometer and support vector machines. *Applied Animal Behaviour Science* 119(1-2), 32-38.

Mattachini, G., Riva, E., Perazzolo, F., Naldi, E. and Provolo, G. 2016. Monitoring feeding behaviour of dairy cows using accelerometers. *Journal of Agricultural Engineering* 47(1), 54-58.

McCulloch, S. P. 2013. A critique of FAWC's five freedoms as a framework for the analysis of animal welfare. *Journal of Agricultural and Environmental Ethics* 26(5), 959-975.

Medrano-Galarza, C., LeBlanc, S. J., Devries, T. J., Jones-Bitton, A., Rushen, J., De Passille, A. M., Endres, M. I. and Haley, D. B. 2018. Effect of age of introduction to an automated milk feeder on calf learning and performance and labor requirements. *Journal of Dairy Science* 101(10), 9371-9384.

Mellor, D. J. 2015. Positive animal welfare states and reference standards for welfare assessment. *New Zealand Veterinary Journal* 63(1), 17-23.

Mellor, D. J. 2017. Operational details of the five domains model and its key applications to the assessment and management of animal welfare. *Animals: An Open Access Journal from MDPI* 7(8), 60.

Mench, J. A. 2003. Assessing animal welfare at the farm and group level: a United States perspective. *Animal Welfare* 12, 493-503.

Molfino, J., Clark, C. E. F., Kerrisk, K. L. and Garcia, S. C. 2017. Evaluation of an activity and rumination monitor in dairy cattle grazing two types of forages. *Animal Production Science* 57(7), 1557-1562.

Morota, G., Ventura, R. V., Silva, F. F., Koyama, M. and Fernando, S. C. 2018. Big data analytics and precision animal agriculture symposium: machine learning and data mining advance predictive big data analysis in precision animal agriculture. *Journal of Animal Science* 96(4), 1540-1550.

Mottram, T. 1997. Automatic monitoring of the health and metabolic status of dairy cows. *Livestock Production Science* 48(3), 209-217.

Mottram, T. T., Dobbelaar, P., Schukken, Y. H., Hobbs, P. J. and Bartlett, P. N. 1999. An experiment to determine the feasibility of automatically detecting hyperketonaemia in dairy cows. *Livestock Production Science* 61(1), 7-11.

Mullins, I. L., Truman, C. M., Campler, M. R., Bewley, J. M. and Costa, J. H. C. 2019. Validation of a commercial automated body condition scoring system on a commercial dairy farm. *Animals: An Open Access Journal from MDPI* 9(6), 287.

Nadimi, E. S., Jorgensen, R. N., Blanes-Vidal, V. and Christensen, S. 2012. Monitoring and classifying animal behavior using ZigBee-based mobile ad hoc wireless sensor networks and artificial neural networks. *Computers and Electronics in Agriculture* 82, 44-54.

Nechanitzky, K., Starke, A., Vidondo, B., Muller, H., Reckardt, M., Friedli, K. and Steiner, A. 2016. Analysis of behavioral changes in dairy cows associated with claw horn lesions. *Journal of Dairy Science* 99(4), 2904-2914.

Nielsen, N. I., Friggens, N. C., Chagunda, M. G. G. and Ingvartsen, K. L. 2005. Predicting risk of ketosis in dairy cows using in-line measurements of beta-hydroxybutyrate: a biological model. *Journal of Dairy Science* 88(7), 2441-2453.

O'Connor, A. H., Bokkers, E. A. M., De Boer, I. J. M., Hogeveen, H., Sayers, R., Byrne, N., Ruelle, E. and Shalloo, L. 2019. Associating cow characteristics with mobility scores in pasture-based dairy cows. *Journal of Dairy Science* 102(9), 8332-8342.

Oudshoorn, F. W., Cornou, C., Hellwing, A. L. F., Hansen, H. H., Munksgaard, L., Lund, P. and Kristensen, T. 2013. Estimation of grass intake on pasture for dairy cows using tightly and loosely mounted di- and tri-axial accelerometers combined with bite count. *Computers and Electronics in Agriculture* 99, 227-235.

Overton, T. R., Mcart, J. A. A. and Nydam, D. V. 2017. A 100-year Review: metabolic health indicators and management of dairy cattle. *Journal of Dairy Science* 100(12), 10398-10417.

Pastell, M., Frondelius, L., Jarvinen, M. and Backman, J. 2018. Filtering methods to improve the accuracy of indoor positioning data for dairy cows. *Biosystems Engineering* 169, 22-31.

Pastell, M., Hautala, M., Poikalainen, V., Praks, J., Veermae, I., Kujala, M. and Ahokas, J. 2008. Automatic observation of cow leg health using load sensors. *Computers and Electronics in Agriculture* 62(1), 48-53.

Pastell, M., Tiusanen, J., Hakojarvi, M. and Hanninen, L. 2009. A wireless accelerometer system with wavelet analysis for assessing lameness in cattle. *Biosystems Engineering* 104(4), 545-551.

Pereira, G. M., Heins, B. J., O'Brien, B., Mcdonagh, A., Lidauer, L. and Kickinger, F. 2020. Validation of an ear tag-based accelerometer system for detecting grazing behaviour of dairy cows. *Journal of Dairy Science* 103(4), 3529-3544.

Poulopoulou, I., Lambertz, C. and Gauly, M. 2019. Are automated sensors a reliable tool to estimate behavioural activities in grazing beef cattle? *Applied Animal Behaviour Science* 216, 1-5.

Reiter, S., Sattlecker, G., Lidauer, L., Kickinger, F., Ohlschuster, M., Auer, W., Schweinzer, V., Klein-Jobstl, D., Drillich, M. and Iwersen, M. 2018. Evaluation of an ear-tag-based accelerometer for monitoring rumination in dairy cows. *Journal of Dairy Science* 101(4), 3398-3411.

Robert, B. D., White, B. J., Renter, D. G. and Larson, R. L. 2011. Determination of lying behavior patterns in healthy beef cattle by use of wireless accelerometers. *American Journal of Veterinary Research* 72(4), 467-473.

Rushen, J., Chapinal, N. and De Passille, A. M. 2012. Automated monitoring of behavioural-based animal welfare indicators. *Animal Welfare* 21(3), 339-350.

Rutten, C. J., Velthuis, A. G. J., Steeneveld, W. and Hogeveen, H. 2013. Invited review: sensors to support health management on dairy farms. *Journal of Dairy Science* 96(4), 1928-1952.

Rutter, S. M. 2014. Smart technologies for detecting animal welfare status and delivering health remedies for rangeland systems. *Revue Scientifique et Technique* 33(1), 181-187.

Ruuska, S., Kajava, S., Mughal, M., Zehner, N. and Mononen, J. 2016. Validation of a pressure sensor-based system for measuring eating, rumination and drinking behaviour of dairy cattle. *Applied Animal Behaviour Science* 174, 19-23.

Schirmann, K., Von Keyserlingk, M. A. G., Weary, D. M., Veira, D. M. and Heuwieser, W. 2009. Technical note: validation of a system for monitoring rumination in dairy cows. *Journal of Dairy Science* 92(12), 6052-6055.

Shalloo, L., O' Donovan, M., Leso, L., Werner, J., Ruelle, E., Geoghegan, A., Delaby, L. and O'Leary, N. 2018. Review: grass-based dairy systems, data and precision technologies. *Animal: An International Journal of Animal Bioscience* 12(s2), s262-s271.

Stangaferro, M. L., Wijma, R., Caixeta, L. S., Al-Abri, M. A. and Giordano, J. O. 2016. Use of rumination and activity monitoring for the identification of dairy cows with health disorders: Part I. Metabolic and digestive disorders. *Journal of Dairy Science* 99(9), 7395-7410.

Steeneveld, W. and Hogeveen, H. 2015. Characterization of Dutch dairy farms using sensor systems for cow management. *Journal of Dairy Science* 98(1), 709-717.

Stone, A. E. 2020. Symposium review: the most important factors affecting adoption of precision dairy monitoring technologies. *Journal of Dairy Science* 103(6), 5740-5745.

Sturm, V., Efrosinin, D., Ohlschuster, M., Gusterer, E., Drillich, M. and Iwersen, M. 2020. Combination of sensor data and health monitoring for early detection of subclinical ketosis in dairy cows. *Sensors* 20(5), 1484.

Svensson, C. and Jensen, M. B. 2007. Short communication: identification of diseased calves by use of data from automatic milk feeders. *Journal of Dairy Science* 90(2), 994-997.

Swartz, T. H., Findlay, A. N. and Petersson-Wolfe, C. S. 2017. Short communication: automated detection of behavioral changes from respiratory disease in pre-weaned calves. *Journal of Dairy Science* 100(11), 9273-9278.

Tamura, T., Okubo, Y., Deguchi, Y., Koshikawa, S., Takahashi, M., Chida, Y. and Okada, K. 2019. Dairy cattle behavior classifications based on decision tree learning using 3-axis neck-mounted accelerometers. *Animal Science Journal = Nihon Chikusan Gakkaiho* 90(4), 589-596.

Thorup, V. M., Munksgaard, L., Robert, P. E., Erhard, H. W., Thomsen, P. T. and Friggens, N. C. 2015. Lameness detection via leg-mounted accelerometers on dairy cows on four commercial farms. *Animal: An International Journal of Animal Bioscience* 9(10), 1704-1712.

Valeeva, N. I., Lam, T. J. G. M. and Hogeveen, H. 2007. Motivation of dairy farmers to improve mastitis management. *Journal of Dairy Science* 90(9), 4466-4477.

Van Asseldonk, M. A., Huirne, R. B. M., Dijkhuizen, A. A., Tomaszewski, M. A. and Harbers, A. G. F. 1998. Effects of information technology on dairy farms in the Netherlands: an empirical analysis of milk production records. *Journal of Dairy Science* 81(10), 2752-2759.

Van de Gucht, T., Saeys, W., Van Weyenberg, S., Lauwers, L., Mertens, K., Vandaele, L., Vangeyte, J. and Van Nuffel, A. 2017. Automatic cow lameness detection with a pressure mat: effects of mat CD length and sensor resolution. *Computers and Electronics in Agriculture* 134, 172-180.

Vandermeulen, J., Bahr, C., Johnston, D., Earley, B., Tullo, E., Fontana, I., Guarino, M., Exadaktylos, V. and Berckmans, D. 2016. Early recognition of bovine respiratory disease in calves using automated continuous monitoring of cough sounds. *Computers and Electronics in Agriculture* 129, 15-26.

Wagner, N., Antoine, V., Mialon, M. M., Lardy, R., Silberberg, M., Koko, J. and Veissier, I. 2020. Machine learning to detect behavioural anomalies in dairy cows under subacute ruminal acidosis. *Computers and Electronics in Agriculture* 170, 105233.

Wang, J., Zhang, H., Ji, J., Zhao, K. and Liu, G. 2019. Development of a wireless measurement system for classifying cow behavior using accelerometer data and location data. *Applied Engineering in Agriculture* 35(2), 135-147.

Webster, J. 2008. *Animal Welfare: Limping towards Eden*. Wiley, U.K.

Werner, J., Umstatter, C., Leso, L., Kennedy, E., Geoghegan, A., Shalloo, L., Schick, M. and O'Brien, B. 2019. Evaluation and application potential of an accelerometer-based collar device for measuring grazing behavior of dairy cows. *Animal: An International Journal of Animal Bioscience* 13(9), 2070–2079.

Williams, L. R., Bishop-Hurley, G. J., Anderson, A. E. and Swain, D. L. 2019. Application of accelerometers to record drinking behaviour of beef cattle. *Animal Production Science* 59(1), 122–132.

Winckler, C. 2019. Assessing animal welfare at the farm level: do we care sufficiently about the individual? *Animal Welfare* 28(1), 77–82.

Zambelis, A., Wolfe, T. and Vasseur, E. 2019. Technical note: validation of an ear-tag accelerometer to identify feeding and activity behaviors of tiestall-housed dairy cattle. *Journal of Dairy Science* 102(5), 4536–4540.

# Chapter 5

## Advances in technologies for monitoring pig welfare

*Maciej Oczak, University of Veterinary Medicine Vienna, Austria; Kristina Maschat, FFoQSI GmbH, Austria; and Johannes Baumgartner, University of Veterinary Medicine Vienna, Austria*

## 1 Introduction

Digital revolution fuelled by the exponential progress of the semiconductor technology has rapidly spread across countries, industries and socio-economic activities in the past few decades, with profound transformational effects (Savulescu 2015, Jorgenson and Vu 2016). In the agricultural industry, especially, areas of food supply chains and arable farming have been influenced by the revolution in the information and communication technology (ICT). This is in contrast to livestock farming, which has been influenced in much lesser extent (Verdouw et al. 2016).

ICT-supported management of livestock farming systems, which is also referred to as precision livestock farming (PLF), was predominantly instigated by the European research community already in the early 1990s (Banhazi et al. 2012). PLF is a tool for the management of livestock farming by means of

http://dx.doi.org/10.19103/AS.2020.0081.15

automatic, real-time monitoring/control of livestock production, reproduction, health and welfare of environmental impacts (Berckmans 2013). Technological development and progress have advanced to such an extent that accurate, powerful and affordable tools are now available. These include cameras, microphones, sensors (such as 3D accelerometers (including gyroscopes), temperature sensors, skin conductivity sensors and glucose sensors), wireless communication tools, Internet connections and cloud storage. Modern technology makes it possible to place cameras, microphones and sensors sufficiently close that they can replace the farmers' eyes and ears in monitoring individual animals (Berckmans 2014).

Apart from sensor data, there is currently an abundance of information available to livestock managers, but it is not generally structured in a way that can be applied readily. For example, a survey of producers raising beef from pastures in southern Australia showed that over 400 pieces of information could be relevant for their farms. The information comes from many sources including academic organizations, government advisors, producer magazines, media sources, technical advisers and other producers (Banhazi et al. 2012).

For monitoring and control of livestock production processes, the PLF approach makes use of modern monitoring and control theory. To achieve favourable monitoring and control of such processes, three conditions must be fulfilled (Berckmans 2006a).

The first condition to be fulfilled is that animal variables must be measured continuously and this information is analysed continuously. 'Animal variables' can be very different such as weight, activity, behaviour, drinking and feeding behaviour, feed intake, sound production, physiological variables (body temperature, respiration frequency, blood variables,). What 'continuously' means is depending on the measured variable such as 25 times a second when monitoring on-line animal activity from video images or a sample every day when monitoring animal weight.

A second condition to realise accurate animal monitoring and management is that at every moment a reliable prediction (expectation) must be available on how the animal variables will vary or how the animal will respond to environmental changes. By environment, we mean the whole of all variables that are not genetically defined. It is the continuous comparison between this prediction (in the past the experience of the farmer and now for example a mathematical model) and the actual measured values that allows to identify animal activities and to judge when something abnormal is happening.

The third condition is that this prediction together with the on-line measurements are integrated in an analysing algorithm (a number of mathematical equations implemented in a microchip) to monitor or manage the animals automatically and to achieve on-line monitoring of animal health, welfare or take control actions (climate control, feeding strategies). A schematic overview of the three conditions is shown in Fig. 1.

The PLF can greatly contribute to an objective discussion on animal welfare by providing real data to the otherwise very subjective (and sometimes emotional) discussion process. While PLF will not be able to necessarily resolve all welfare-related questions, it will allow interested parties to detect and act upon time periods when animals are kept under suboptimal conditions (Banhazi et al. 2012).

Today, early adopters are starting to use PLF products mainly in Europe, despite the PLF concept is rather new in the European pig industry. So far, there is a lack of products in the market, which would comprehensively approach monitoring the welfare of pigs. However, there are already systems available which allow monitoring individual indicators of pig welfare such as water intake or weight (Kashiha et al. 2013a, Vranken and Berckmans 2017). There are at least two major advantages of PLF technology, which will lead to its wider adoption for welfare monitoring in the coming years.

One advantage of sensor-based monitoring is its continuous character (Berckmans 2006a). This means that monitoring can be always on 7 days a week, 24 h per day. This is clearly advantageous if we compare this to the time that the farmer can directly observe animals in industrial husbandry systems. Thus, continuous monitoring can create a possibility to bring the animals closer to the farmer, especially in industrial systems, where contact between an individual

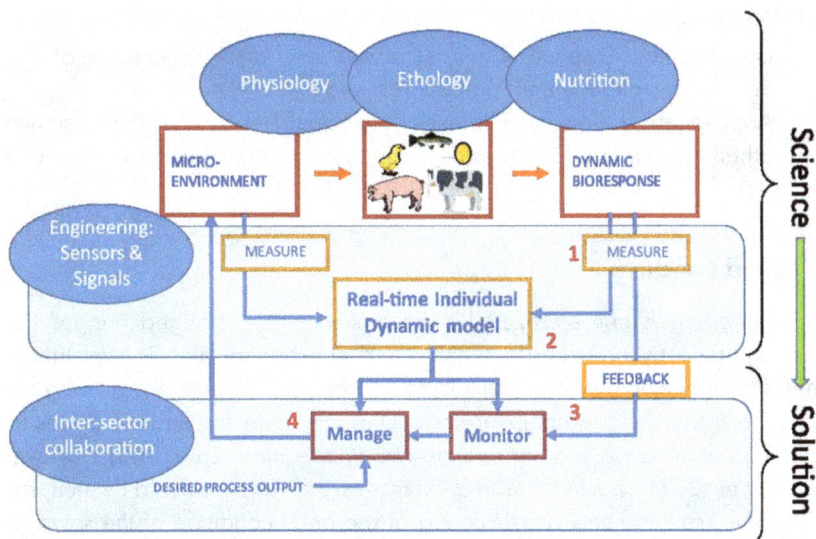

**Figure 1** General scheme showing how bio-response monitoring and management of livestock animals can go from science to solutions. From Aerts, Norton, and Berckmans 2019.

animal and the farmer is very limited. The farmer can obtain insight about the animal and its welfare, thanks to sensor technology, which is not possible on the basis of short-lasting, personal audio-visual observation. Thus, the farmer can invest his time in the animals that require his attention.

The second advantage of PLF systems is that they offer real-time monitoring and this gives a possibility to respond to welfare problems immediately, when the problem occurs. This is fundamentally different from all other approaches that aim to offer a monitoring tool without improving the life of the animal under consideration. It is nice to detect a problem after an animal has arrived at the slaughterhouse. It is better to detect a problem while the animal is being reared and to take immediate management action (Berckmans 2013). However, implementation of automated monitoring technologies at the abattoir, while less immediate in terms of problem identification, may be much more cost-effective in some circumstances, for example, routine disease or welfare surveillance programmes.

In this chapter we divided advances in the development of PLF technologies for monitoring welfare in pigs into four main parts according to the grouping of welfare criteria based on four principles proposed in Welfare Quality® (WelfareQuality® 2009):

- Good feeding: Absence of prolonged hunger; absence of prolonged thirst;
- Good housing: Comfort around resting; thermal comfort; ease of movement;
- Good health: Absence of injuries; absence of disease; absence of pain induced by management procedures; and
- Appropriate behaviour: Expression of social behaviours; Expression of other behaviours; Good human-animal relationship; Positive emotional state.

## 2 Good feeding

Good feeding is characterized by the absence of prolonged hunger that is guaranteed by providing a suitable and appropriate diet. The evaluation method suggested by WelfareQuality® (2009) is visual inspection (growing and finishing pigs, sows) and palpation (sows) of the spine, hip and pin bones for thickness of fat tissue in order to assess body condition score (BCS). However, Maes et al. (2004) showed that back fat measurements conducted by means of ultrasonic scanning and visual scoring of the body condition of the sows are only moderately correlated (r=0.30-0.60). Although depth cameras measuring back postures have been used lately (Kuzuhara et al. 2015, Weber et al. 2014) in cows to predict body condition score and back fat, this method hasn't been adapted to pigs so far. Yet, depth image and machine learning have been applied to estimate muscle scores of in vivo pigs (Alsahaf et al. 2019).

Also, it has to be kept in mind that decreases in BCS in the context of malnutrition and diseases occur at a very late stage when the course has already become chronic. Therefore objective assessment of other indicators for feeding might be more beneficial. Automatic weight detection has been an approach to address this problem and at the same time bears the potential to increase precision of marketing pigs (Condotta et al. 2018). Recent studies have been using pigs' body dimensions assessed on the basis of top-view 2D- and 3D-camera imaging for predicting animals' mass.

Some of the authors separated pigs from their group to conduct measurements while the animal was confined (Condotta et al. 2018, Kongsro 2014) and reached very high coefficients of determination ($R^2$=0.99), while others have been aiming to develop non-contact methods to estimate the weight of pigs in barns with other animals (Jun et al. 2018, Pezzuolo et al. 2018, Kashiha et al. 2014) and achieved results with $R^2$ ranging from 0.79 to 0.96. Mobile devices have been developed (Shi et al. 2016, Wang et al. 2018) to estimate pig body sizes and commercial products for weight estimation such as Weight-Detect (PLF-Agritech Europe, Edinburgh, UK), Pigwei (Ymaging, Barcelona, Spain), eYeScan (Fig. 2) (Fancom BV, Panningen, The Netherlands), Growth Sensor (GroStat, Newport, UK), OptiSCAN (Hölscher, Emsburen, Germany), and WUGGL One (WUGGL, Lebring, Austria) have been introduced to the market (Vranken and Berckmans 2017).

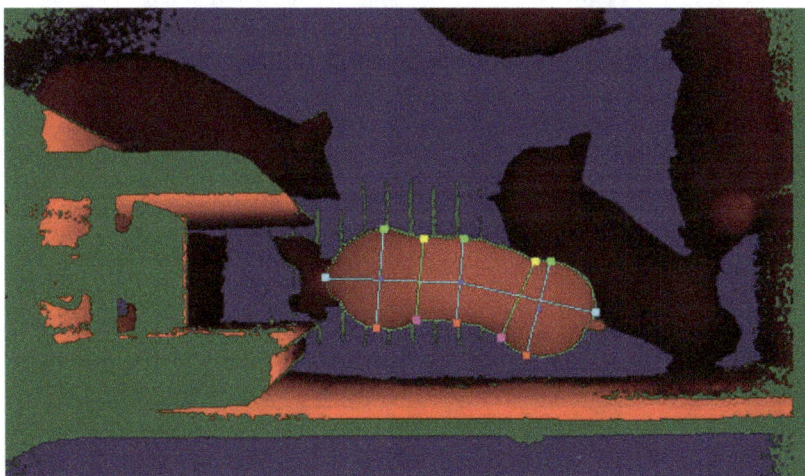

**Figure 2** Image of eYeScan (3D) showing the step in the process where the body characteristics are determined to calculate the pig features such as body width, length and area. From Vranken and Berckmans 2017.

Precision feeding techniques use real-time measurements such as body weight and feed intake to provide each animal with the right amount and composition of feed at the right time (Pomar and Remus 2019, Pomar et al. 2009). They therefore can improve production strategies, meat-product quality as well as animal welfare and also have the potential to reduce environmental footprint of pig production systems. Nitrogen and phosphorous excretion (Andretta et al. 2014, Andretta et al. 2016) lysine intake (Andretta et al. 2016) and greenhouse gases emissions (Andretta et al. 2018) can be significantly reduced by increasing individual nutrient efficiency (as reviewed by Pomar and Remus (2019)).

An important feature for precision feeding is individual pig identification that is usually provided by radio-frequency identification (RFID) technology. RFID technology uses electromagnetic fields to automatically identify and track either active (with a local power source) or passive (receiving energy from an RFID reader) tags attached to objects. Its application is well known in farm animal husbandry in the context of restricted feeding, where a passive transponder is mounted to the ear of the animal and a stationary reader registers the animal's visits.

For on-farm animal management purposes low-frequency RFID systems are common as they offer high reading reliability (Brown-Brandl et al. 2017). However, they are unable to register multiple transponders simultaneously (Adrion et al. 2018) in contrast to high- and ultra-high frequency RFID systems that apply an anti-collision protocol and have higher data rates and read ranges (up to 12 m for ultra-high frequency) (Brown-Brandl et al. 2017). Recent publications in the field of animal behaviour have mainly been using high as well as ultra-high frequency RFID systems to assess feeding patterns on the basis of individual frequency and duration of feeding place visits in groups of pigs of different production stages (Maselyne et al. 2016b, Maselyne et al. 2016a, Maselyne et al. 2014, Reiners et al. 2009, Adrion et al. 2018). An online warning system for individual fattening pigs on the basis of feeding patterns determined by RFID data has been developed (Maselyne et al. 2017).

Water supply and intake can be tested and measured at group-level by a water flow-meter as a stand-alone device. Recent studies have combined the water flow-meter with RFID systems to examine drinking behaviour in individuals (Maselyne et al. 2016a, Andersen et al. 2014). Madsen et al. (2005) successfully modelled the water consumption of young growing pigs on the basis of flow-meter data by combining a growth model and a cyclic model in order to predict their drinking behaviour. However, water wastage has to be considered as a factor that might influence water intake data on the basis of water flow measurements. Kashiha et al. (2013b) applied automatic image processing based on top-view video recordings and transfer function modelling to estimate water use and reached an accuracy of 92%.

## 3 Good housing

Thermal comfort of pigs as one of three criteria of welfare within a principle of good housing (WelfareQuality® 2009) was traditionally monitored with sensor technology based on predetermined ambient temperature on a compartment level. Advancement in sensor technology allows placing thermometers in ear tags and measures body temperature directly on the individual animal (Andersen et al. 2008). The results of this experiment indicate that pigs adjust their behaviour to a higher ear surface temperature and they use behavioural adjustment (e.g. increased/decreased contact to pen mates) to bring their skin temperature into a preferred interval (Andersen et al. 2008).

The traditional approach of estimating thermal comfort of pigs can fall short in meeting the animals' true thermal needs because it does not integrate the effects of other contributing factors, such as drafts, humidity (particularly in hot conditions), radiation (in poorly insulated barns), floor type and/or condition (dry vs. wet, bedding vs. no bedding), nutritional plane and health status of the animal (Shao and Xin 2008).

The best indicator of the environment adequacy and thus animal comfort is animals themselves that integrate both external and internal factors, which in turn lead to distinctive resting behaviours. Huddling, resting next to one another, and spreading are the stereotypical postural patterns of group-housed animals when experiencing cold, comfortable, and warm/hot sensation, respectively (Shao and Xin 2008).

For monitoring pig behaviour as an indicator of thermal comfort 2D camera sensors were used in the research of Costa et al. (2014), Shao and Xin (2008) and Xin (1999). Costa et al. (2014) monitored behaviour of a group of fattening pigs on the basis of two variables extracted from images, the activity and occupation indexes. It was possible to automatically detect that pigs spent less time in the area of the pens with a humid floor surface and with a limited air speed. The animals occupied 'Central zone' in the barn, which was characterized by dry floor and faster air movement, more often.

In the research by Xin (1999) and Shao and Xin (2008), image analysis was used to classify thermal comfort state of group-housed pigs on the basis of their resting behaviour. The human observer first labelled images with pigs' resting behaviour as cold, comfortable and warm sensations. In the next steps images were segmented, motion detection and feature extraction was applied. Finally, the thermal comfort state was automatically classified. The test results showed over 90% of correct classification rate. This result suggests that continuous, automated classification of thermal comfort in fattening pigs measured directly on the animal is possible with modern image analysis techniques.

Image analysis was also applied in research of Nasirahmadi et al. (2017) to classify group-lying patterns of pigs (close, normal and far) in different thermal categories in commercial pig-farm conditions. An image processing technique based on Delaunay triangulation (DT) was utilized. The percentage of each defined lying pattern, obtained through calculating the perimeter of each triangle in the DT, changed significantly as the environmental temperatures increased. The authors conclude that the proposed method is an important step towards improving animal welfare in commercial farm conditions with their changeable environmental parameters.

Ease of movement was traditionally evaluated by average space available per animal in a pen. Different recommendations for appropriate space allowance were given depending on study design from which they were derived. A combination of three-dimensional cameras and image analysis allows automated measurement of space used by the animal and free space available in a pen (Fels et al. 2018). This opens up possibilities not only to assess spatial requirements for pigs in a new way but also to measure it in real-time as animals grow and take up more space in a pen (Fig. 3). Additionally, there is potential in the application of image analysis for monitoring the ease of movement of pigs by tracking the distance moved by an individual in a group (Matthews et al. 2017).

## 4 Good health

Animals in good health should be free of injuries (e.g. skin damage and locomotory disorders) and from disease (i.e. animal unit managers should

**Figure 3** Image of a piglet group generated by the programme for image analysis. Animals identified as individuals have a green line drawn around them and are numbered 1–3. From Fels et al. 2018.

maintain high standards of hygiene and care) and should not suffer pain induced by inappropriate management, handling, slaughter or surgical procedures (WelfareQuality® 2009).

Lameness has a major impact on health and welfare of pigs and is of high economic importance, especially in sows where it is a major cause of culling (as reviewed by Heinonen et al. (2013)). Lame sows spend more time lying and less time moving and standing than healthy ones (Ala-Kurikka et al. 2017). Their stepping behaviour is increased (Conte et al. 2015), they walk more slowly, have a shorter stride length, a longer stance time and a shorter latency to lie down after feeding (Gregoire et al. 2013). So far, technologies as force plates (Conte et al. 2014, Pluym et al. 2013, Karriker et al. 2013, Mohling et al. 2014), pressure mats (Meijer et al. 2014), infrared thermography (Amezcua et al. 2014), accelerometers (Gregoire et al. 2013, Conte et al. 2015, Traulsen et al. 2016, Scheel et al. 2017), electronic feeders (Cornou et al. 2008) and image analysis (Stavrakakis et al. 2015, Pluym et al. 2013) have been applied to replace visual examination in lameness detection in pigs.

Camera-based, automated systems for the assessment of lesions have been developed for ear and tail lesions at the abattoir (Blömke et al. 2020, Brünger et al. 2019). Blömke et al. (2020) detected ear lesions with an accuracy of 95.4% and for tail lesions an accuracy of 99.5% was achieved. A future vision is automatic screening of pathological organ alterations. Initial steps towards this direction have already been taken. McKenna et al. (2018) tested a method based on auto-context segmentation on images of porcine offal in an abattoir production line. Highest dice coefficients for different organs ranged between 0.816 (heart) and 0.973 (liver).

Many studies have been focussing on automated monitoring and early detection of diseases. Munsterhjelm et al. (2015) used automatic single-space feeders in growing pigs and found a decrease of feed intake in future lame or tail-bitten individuals already 2-3 weeks before diagnosis. Electronic sow feeders have been evaluated to detect estrus, health as well as pregnancy disorders in group-housed sows (Cornou et al. 2008, Iida et al. 2017). Madsen and Kristensen (2005) applied time-series analysis on water flow-data in groups of growing pigs and were able to detect increasing water intake caused by diarrhoea 1 day before symptoms were recognised by the care-taker. Another approach to monitor disease in pig herds was developed by Martínez-Avilés et al. (2015). Pigs were infected with an attenuated strain of African swine fever and equipped with acceleration, body temperature and RFID sensors to monitor movement, body temperature and water intake. Additionally, cameras were collecting data for video image processing. This real-time monitoring system was able to identify the infection 1-3 days before qPCR detection.

Recently, infrared thermography has gained attention as a non-invasive remote sensing method that has been used to indicate thermal biometric changes in animal metabolism (McManus et al. 2016). When assessed correctly, the measured parameter has a high correlation with body temperature (as reviewed by Soerensen and Pedersen (2015)) and can be used as a feature variable, for example, for febrile responses (Cook et al. 2015) and hypothermia in newly born piglets (Kammersgaard et al. 2013). Lu et al. (2018) were able to measure ear base temperature on the basis of top-view piglet thermal images with an accuracy of 97-98%.

Most of the above-mentioned methods for disease monitoring and early detection are sensitive enough to determine a significant change in indicator values. However, most indicators are not pathognomonic and therefore allow only vague assumptions on the cause of the change. Automated monitoring of vocalizations associated with respiratory diseases permits a restriction of tentative diagnoses to those with manifestation in the respiratory tract. Audio surveillance systems that record coughing do not only allow identification of sick pig cough sounds in real time (Exadaktylos et al. 2008), but also differentiation between non-infectious and infectious coughs with further characterization of coughs caused by pathogens such as Actinobacillus pleuropneumoniae and

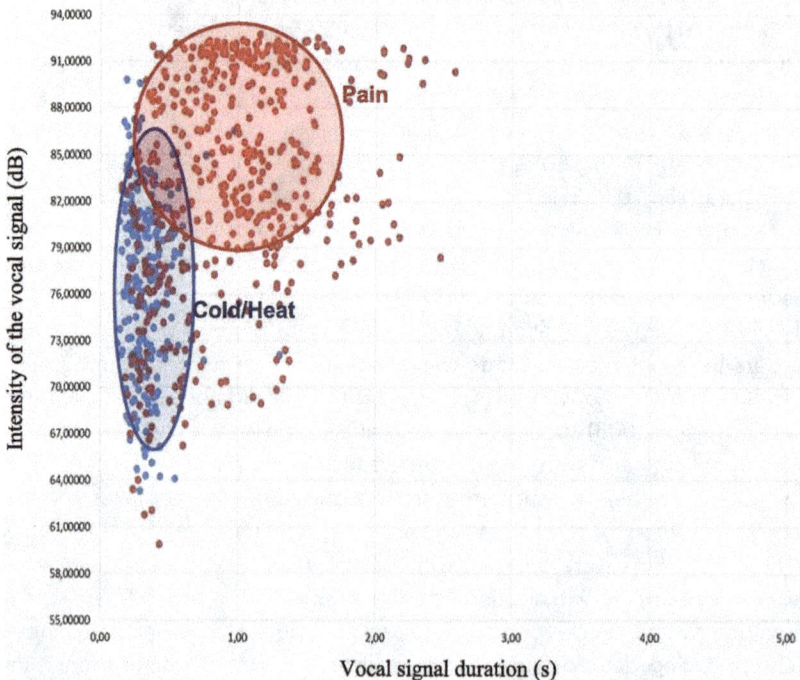

**Figure 4** The output of the software using the sound wave intensity (vertical axis, dB) and the duration of the vocal signal (horizontal axis, s) during the exposition to heat and pain. From da Silva et al. (2019).

Pasteurella multocida (Ferrari et al. 2008). Chung et al. (2013) were able to make a distinction based on cough sounds between postweaning multisystemic wasting syndrome, porcine reproductive and respiratory syndrome and mycoplasma hyopneumoniae. Cough sound analysis has also been applied to assess air quality in weaner barns (Wang et al. 2019).

Moreover, sound data can also be helpful in the assessment of pain. Experiencing different stages of pain has been associated with differing vocalization patterns (Leidig et al. 2009, Taylor and Weary 2000, Weary et al. 1998, Diana et al. 2019). Pain was defined as a target variable in research from da Silva et al. (2019), who were able to predict pain on the basis of vocal intensity and duration in piglets with an accuracy of 93.0% (Fig. 4).

## 5 Appropriate behaviour

According to the WelfareQuality® (2009) principle, 'appropriate behaviour' animals should be able to express normal, non-harmful, social behaviours (e.g. grooming) and other normal behaviours (i.e. foraging or exploring). They should be handled well in all situations, that is, handlers should promote good human-animal relationships and negative emotions such as fear, distress, frustration or apathy should be avoided whereas positive emotions such as security or contentment should be promoted.

One of the few published attempts to automatize the detection of social behaviours was made by Šustr et al. (2001). They used the EthoVision® tracking system (Noldus Information Technology, Wageningen, the Netherlands) to provide coordinates of colour-marked body parts of two pigs and then applied an algorithm to determine whether these two pigs were in body contact, their mutual orientation and whether the pig in the 'active' position made snout contact in front of or behind the ear base of the other pig. This method might also be useful for automatic detection of play behaviour, a motivated behaviour that only occurs under currently non-life-threatening conditions (Held and Špinka 2011) and is considered as an indicator for positive welfare (Lawrence et al. 2018, Boissy et al. 2007). Other studies have been focussing on object-oriented play behaviour by equipping the object with a sensor (Zonderland et al. 2003). Another interesting approach to investigate social interactions and at the same time open new doors to a new dimension of 'social engineering' in pigs was developed by Boumans et al. (2018). They designed an in-depth agent-based model to simulate social interactions and feeding behaviour of individually and group-housed pigs and suggested the implementation of this model in research as a tool for studying the effects of social factors and group dynamics on individual variation in feeding and social interaction patterns.

Automatic detection of mounting, a behaviour that occurs during the entire growing period (Thomsen et al. 2012), but more frequently in overcrowded situations (Faucitano 2001) has been studied by Nasirahmadi et al. (2016). They achieved an accuracy of 92.7% by applying image analysis with an ellipse fitting technique.

Rooting activity as part of exploratory behaviour has been automatically detected on the basis of 3D accelerometer data with an accuracy ranging from 66.5% (Escalante et al. 2013) to 86.7% (Cornou and Lundbye-Christensen 2008). For the purpose of automatic detection of nest-building Oczak et al. (2015) observed pawing, exploratory behaviour, manipulation of pen and rack. The accuracy they found for 'nest-building activity' as combination of those four behaviours was 86%.

As the awareness and demand for assessment of positive welfare is relatively new (as reviewed by Yeates and Main (2008) and Lawrence et al. (2018)), PLF monitoring attempts have been mainly focussing on potentially harmful behaviours like aggression as welfare indicators. Although low-level aggression is considered as very important for the maintenance of a stable social hierarchy in pig groups (McGlone 1985), high occurrence (usually linked to mixing unfamiliar pigs) is associated with stress and reduced animal welfare (McGlone et al. 1980, Turner et al. 2006).

Image feature extraction analysis has been applied by Viazzi et al. (2014) to detect aggression automatically while Chen et al. (2019) combined this method with a depth sensor reaching an accuracy of 97.5%. Oczak et al. (2014) proposed a technique based on an artificial neural network that classified high and medium aggression from video recordings. Chen et al. (2018) were able to do the same classification using a kinetic energy model based on machine vision.

Abnormal behaviour can originate from unsuccessfully applied coping strategies in aversive, overstraining situations (Wechsler 1995). Tail biting often occurs in situations when environmental factors do not comply with animals' needs (Hunter et al. 2001, Moinard et al. 2003, van Putten 1969) and the animals fail to gain control over this situation. It is considered as a welfare-reducing problem (European Food Safety Authority 2007) with economic consequences for pig production.

Former studies have proved indicators such as increased pig (Statham et al. 2009, Ursinus et al. 2014, Zonderland et al. 2011) and pen (Ursinus et al. 2014) directed manipulative behaviours, object manipulation (Larsen et al. 2019a), amount of posture changes (Zonderland et al. 2011), activity (Larsen et al. 2019a, Statham et al. 2009, Ursinus et al. 2014, Zonderland et al. 2011), hanging tail posture (Lahrmann et al. 2018, Wedin et al. 2018) and low feed intake (Wallenbeck and Keeling 2013) as reliable predictors for an outbreak. Yet, only few studies have been linking these behavioural parameters to PLF technologies. Larsen et al. (2019b) collected sensor data on water usage and pen temperature to develop an algorithm for tail biting while D'Eath et al.

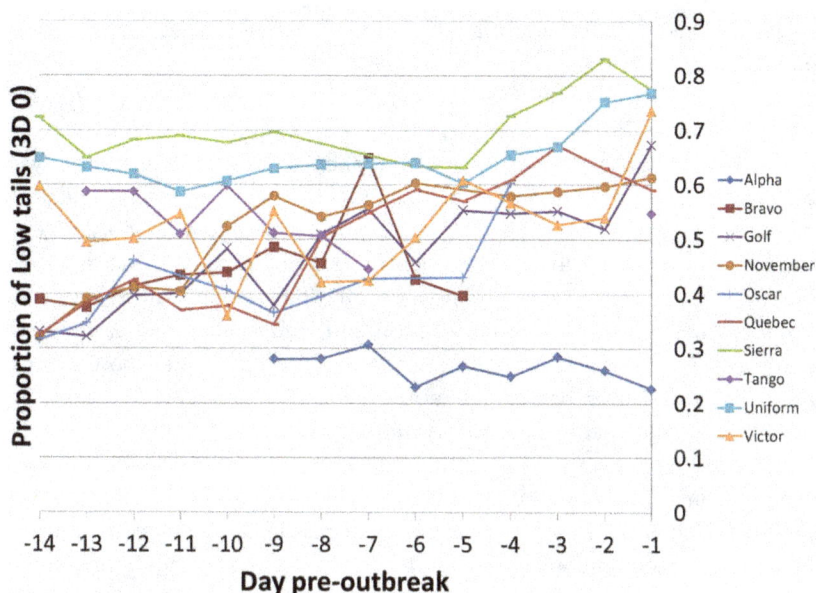

**Figure 5** Proportion of 3D low tail detections (3D 0) on the days leading up to an outbreak. Data are shown for the 10 outbreak groups. Each line shows data for a different group. From D'Eath et al. (2018).

(2018) were able to detect lowered tail posture automatically on the basis of 3D-cameras before the outbreak (Fig. 5). As tail (and ear) posture and movement have been found to be indicative for emotions in pigs (Reimert et al. 2013, Marcet Rius et al. 2018), the latter method might have much potential for future applications in the context of monitoring welfare.

Another promising tool to monitor appropriate behaviour as well as emotions of pigs that has already been mentioned in the subchapter 'Good health' is audio surveillance of vocalizations (as reviewed by Manteuffel et al. (2004)). Automation techniques have been so far mainly focussing on stress responses (Schon et al. 2004, da Silva et al. 2019, Moura et al. 2008).

## 6 Legal and ethical considerations

In PLF systems, mainly non-invasive sensors such as cameras and microphones are used. If sensors have to be placed on (i.e. ear) or in the animal (i.e. rumen of cattle), this may under no circumstances result in significant additional pain, inflammation, functional restriction or any other risk of impaired welfare of the animal. In this regard, active sensors powered by long-life batteries can cause problems due to weight and size. In any case, the attachment of PLF sensors on or in the animal must comply with the respective animal welfare regulations

(European Union 2008b) and the animal identification regulations (European Union 2008a).

PLF-based information systems must be developed according to strict scientific criteria. The selection of suitable welfare indicators, the definition of the relevant gold standards, the assurance of reliability and validity in primary data processing and the definition of standard ranges must be based on comprehensible physiological, ethological and veterinary criteria. If inaccuracies occur at this early stage of the developmental process of PLF systems, it may have a negative impact on the health and well-being of the monitored animals due to misinformation and inadequate reaction. In order to avoid situations where poorly designed PLF products enter the market, it is recommended to consider the regulation of PLF products following the example of human medical products (European Union 2017).

One way of viewing PLF is to see it as a ground for improving the well-being of animals, because the technology highlights the care for the individual animals and their quality of life. On the other hand PLF mediates the human-animal relationship in conventional industrial livestock farming, implying that farmers and animals become subjected to new expectations. PLF redefines the notion of care, in terms of data transparency, standardization of methods for analysis, real-time collection and processing of data, remote control, and the use of digital platforms. Therefore PLF requires a redistribution of responsibilities within a wider scope of relations in the value chain (Bos et al. 2018).

PLF not only has a direct impact on animal welfare by improving animal monitoring, facilitating standardized decisions and thereby positively affecting animal care. It is also very likely that PLF technology, consisting of biometrics, Internet of Things, robotics, augmented reality and digital platforms will fundamentally change the entire animal food production chain and thereby indirectly affecting animals' quality of life.

By means of algorithms a huge amount of data is transformed in real-time to information on the health and welfare status of individual or groups of animals. Through systems like artificial intelligence, algorithms do not follow a predetermined set of rules but make use of self-learning statistical techniques. As a result, algorithm-based decisions are almost unfathomable and uncontrollable for humans. To prevent manipulation, it is therefore crucial that we understand why such algorithms make certain choices, and how to implement transparency (Royakkers et al. 2018).

From an ethical point of view, the question arises as to whether the dynamics of digitalization are changing our understanding of the subjectivity of humans and animals in conventional industrial livestock farming. According to Bos et al. (2018), the transformative effects of PLF on the human-animal relationships turn animals and human into objects. Royakkers et al. (2018) describe effects such as deanimalization, dehumanization, instrumentalization, desocialization and deskilling to those jobs that remain. However, this development has been observed in industrial livestock farming for decades.

PLF may primarily be used to enhance the viability of intensive livestock production, to make it more feasible to keep very large herds in stressful, high-density conditions with poor levels of welfare. By alerting farmers to problems at an early stage it can, to a degree, improve animal welfare and system efficiency. However, such improvements are made within a system that has inherently low potential for good welfare (Stevenson 2017). Another issue is the further consolidation of farms, as only those with the capital to invest in PLF can benefit from the 'technology treadmill' of ever-improving PLF technologies (Werkheiser 2018).

Bos et al. (2018) conclude that PLF requires a new balance of power between animal, farmer, veterinarian, industry, consumer and government in order to prevent unfair competition and exploitation. Actors with strong commercial interests are particularly likely to use PLF data to gain market power. If products become more dependent on PLF software, this strengthens PLF-manufacturers' control and how that can be utilized (Royakkers et al. 2018). This may also mean that ignoring animal welfare principles by farmers will no longer be invisible and tolerated.

## 7 Sensor data management and modelling

In the livestock process, the central and most complex component is the animal. Due to the time-varying behaviour of animals being living organisms, the monitoring of the PLF approach requires continuous measurements of animals' responses directly on the animal rather than in the environment surrounding the living organism (Berckmans 2017).

Since animal responses can be very fast, it is useless to carry out a survey once a year, once a month or week, or even twice a day. We need a continuous monitoring/management tool. Depending on the variable that is monitored, the word 'continuous' might mean every second (e.g. for stress monitoring) or once a day (e.g. for weight monitoring). Modern technology makes it all possible. The field data consist of a lot of numbers originating from the sensors (e.g. 240 samples per second for an accelerometer), images (e.g. 25 images per second), or sound signals (e.g. 20 000 samples per second). Within the EU-PLF project (EU-PLF 2016), the PLF technology was installed in 5 broiler houses and 10 pig houses. The monitoring over 90 fattening periods for pigs in 3 years resulted in 5475 measuring days, generating more than 120 terabytes image data and 4 906 000 files of 5 min each with sound data (Berckmans 2017).

When the PLF technologies are installed in livestock houses, they generate a huge amount of data, and the transmission of so much data takes time, energy and money. Sending data wirelessly involves energy and costs; we should, therefore, avoid transmitting too much data and develop real-time algorithms that calculate information from the data at the lowest possible level enabling us to transmit relevant information rather than data. Therefore, we need real-time algorithms that can calculate relevant information from the data preferable on or close to the individual animal (Fig. 6) (Berckmans 2017).

Living organisms are CITD systems, which stand for complex, individually different, time-varying and dynamic systems (Berckmans 2006b). It is obvious than a living organism is much more complex than any mechanical, electronic or ICT system. The complexity of information transmission in a single cell of a living organism is much higher than most other systems that may be considered. The time-varying character of living organisms means that a living organism's

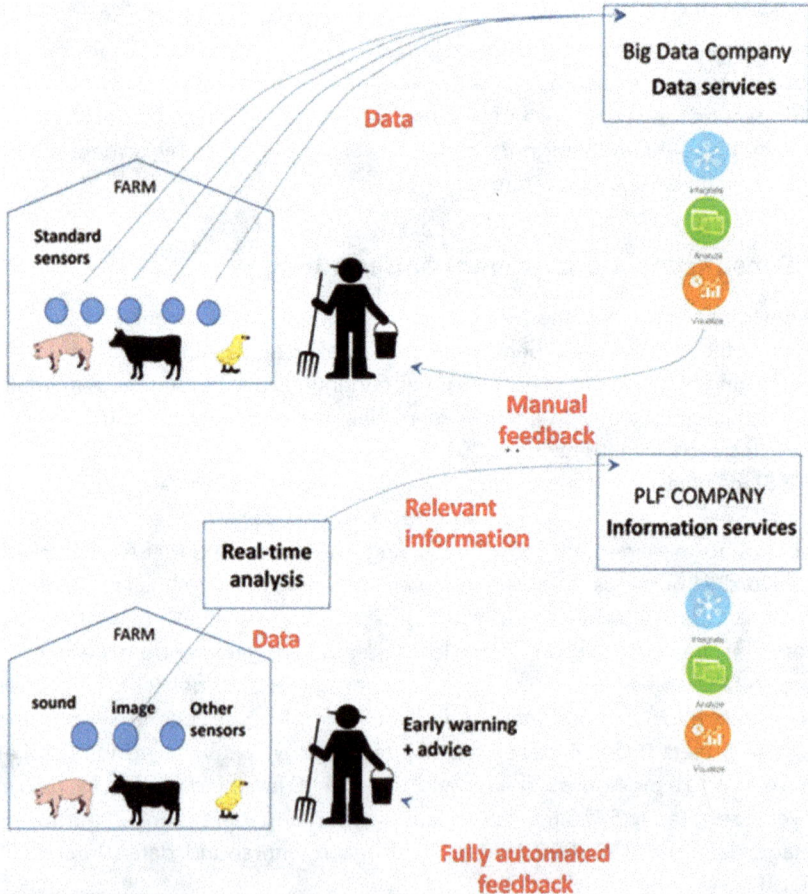

**Figure 6** Scheme of a more realistic use of data versus the common idea of big data. From Berckmans 2017.

response to an environmental stimulus or stressor might be different each time it happens. A living organism is constantly looking for a good energy balance and as a consequence is continuously changing its physical condition and mental status. Of course living organisms are dynamic systems (Berckmans 2013).

The CITD nature of living organisms has an important impact on the type of algorithms that need to be developed for PLF applications. It implies that algorithms to monitor these time-varying individuals must continuously adapt to the individual and/or use principles that can be used in real time in the field application (Berckmans 2013). Algorithms from the family of dynamic models (e.g. dynamic regression, transfer function) are appropriate for modelling CITD systems (Young 2011).

The advantages of these algorithms are that they are simple, suitable for online and real-time monitoring and also are interpretable. Appropriate model order identification criteria such as Akaike (AIC) (Akaike 1974), Bayesian (BIC) (Akaike 1974) or Young Information Criterion (YIC) (Young 1984) are used together with statistical measures of how well models fit to data such as coefficient of determination, $R^2$, to identify models often with only few parameters.

Such simple models can be interpreted meaning that conclusions can be drawn about the underlying modelled biological process in relation to variation in model parameters (Fernández et al. 2019). Simplicity of these models also offers possibility to implement them on embedded systems, on cheap devices, close to the animal, reducing the need to transfer large amounts of sensor data to remote locations. Dynamic models can be estimated for each individual animal. Because model parameters evolve over time these models adapt accordingly to time-varying response of animals. This allows those models to achieve similar, robust performance when validated on different farms (Oczak 2018).

Algorithms from the family of machine learning (support-vector machines, artificial neural networks) that gain popularity in recent years also in the research field of animal monitoring (Nasirahmadi et al. 2017, Valletta et al. 2017, Escalante et al. 2013) are complex (1000s of parameters) and are black box models (Hansen et al. 2018). This means that these models require much more calculations than simple models. Thus, powerful hardware is needed for estimation and implementation of these models for real-time processing of sensor data. Nowadays such powerful hardware can be implemented on farms; however the cost is considerably higher in comparison to hardware cost of implementation of simple models. Black box character of machine-learning algorithms limits the possibility to gain insight on modelled biological processes as variation in 1000s of parameters cannot be related to these processes in comprehensible ways. With these disadvantages of machine-learning methods the main advantage is that very complex sensor data such as image data can be modelled with less effort for the modeller. These models possibly can be a basis for robust monitoring systems in PLF if enough variable input data is used for training (Oczak 2018).

New developments in machine-learning algorithms such as recurrent neural networks (RNN) can be suitable to model CITD systems as RNN can be individually estimated on time series data. Parameters of RNN algorithm similar to dynamic models evolve over time so that these models adapt to time-varying response of individual animals (Chen et al. 2020).

# 8 Case study

It is a common practice in modern intensive pig husbandry to confine sows in farrowing crates, usually for 4–5 weeks, including at least few days before the onset of farrowing. The main reason for this practice is to improve piglet survival rate by protecting new-born piglets from fatal or injurious crushing by the mother sow (King et al. 2019). However, the confinement of sows in crates has a negative impact on sows' welfare, such as limited freedom of movement, limited social interactions with new-born piglets (Melišová et al. 2011, Pedersen et al. 2013) and diminished health (Lambertz et al. 2015, Singh et al. 2017). Confinement also prevents much of the prenatal nest-building behaviour, an essential part of behavioural repertoire in sows, which starts around 24 h before parturition, is most intense 6–12 h before parturition and then decreases as parturition approaches (Wischner et al. 2009, Castrén et al. 1993). Increased physiological stress for the sow is a consequence of confinement in a crate (Jarvis et al. 2006).

A concept of temporary crating has been developed as a response to increased public concern about welfare of crated sows (Moustsen et al. 2012). According to this concept sows should be temporarily confined in farrowing crates only during the critical period of piglets' life, when piglet crushing is most probable, in the first days after farrowing (Marchant et al. 2000, BMGÖ 2012). When the crate is opened farrowing pen offers additional space for the sow, providing a compromise between the needs of the farmer, the sow and her piglets (King et al. 2019). This would allow the sow to stay unconfined during the prenatal nest-building phase from 24 h ante partum until approach of farrowing, which would have a positive impact on sows' welfare (Algers 1994). However, choosing the right moment to confine an individual sow in a farrowing crate under farm conditions in a way that makes nest-building possible and does not increase risk of piglet crushing is challenging. Due to the biological variability in gestation length, time-consuming observation of sows would be necessary. On the other hand, confinement of sows in crates based on a calculated farrowing date of the batch could either disturb adequate nest-building or the farrowing process in many sows.

Automated detection of changes in sow activity, with the use of sensor technology, was used in the research by Oczak et al. (2019) to indicate when sows should be confined in a farrowing crate. Ear tag-based acceleration data was modelled to provide two types of alarms: 'First stage' alarm indicates the beginning of nest-building behaviour while 'second stage' indicates the end of nest-building behaviour. In total, 53 sows were included in the experiment, 27 in the training data set and 26 in the validation data set. Each sow had an ear tag with an accelerometer sensor mounted on the ear. Acceleration data was modelled with Kalman filtering and fixed interval smoothing (KALMSMO) algorithm.

In the first step total physical acceleration (magnitude) was estimated from three axes of accelerometer data (xyz). In the second step total physical acceleration (magnitude) was smoothed with standard deviation calculated on a sliding window of 24 h with 15 min steps (Fig. 7).

Smoothing window of 24 h allowed the elimination of variation in the activity of animal related to diurnal rhythm. This input variable was used to estimate a trend in the activity of each sow. Changes in trends (dynamics) in the activity of sows were a basis for detection of approaching farrowing and providing first and second level alarms. To estimate the dynamics of activity of sows the KALMSMO algorithm was used that comprise as a special case – the Integrated Random Walk (Young 2011).

In the consecutive steps the fixed interval was expended recursively by 15 min steps until the trend in animal activity changed to significantly increasing. This was indicated by the input variable reaching higher value than upper confidence interval of estimated trend (Fig. 8). At this time point the 'first stage' alarm informing about approaching farrowing was raised.

In the next step KALMSMO algorithm was applied to acceleration data on a fixed interval starting from the time point of the 'first level' alarm. Next, the fixed interval was expended recursively by 15 min steps until the trend in animal activity changed to significantly decreasing. This was indicated by an input variable reaching a lower value than the lower confidence interval of the estimated trend (Fig. 9). At this time point the 'second stage' alarm was raised. This alarm could be interpreted as an indication to confine a sow in a crate.

It was possible to predict farrowing on the basis of increased activity in the validation dataset with a median of 8 h 51 min before the onset of farrowing. Alarms indicating the need for confinement of the sow in a crate were raised with a median of 2 h 3 min before the onset of farrowing (Fig. 10).

Practical application of 'first stage' alarms generated around 6-13 h before the onset of farrowing might be to warn the farmer about approaching farrowing in an automated way. This could reduce labour costs otherwise required for regular control of sows in farrowing compartments. It should also limit the need to get in close contact with sows which could interrupt their activities at a

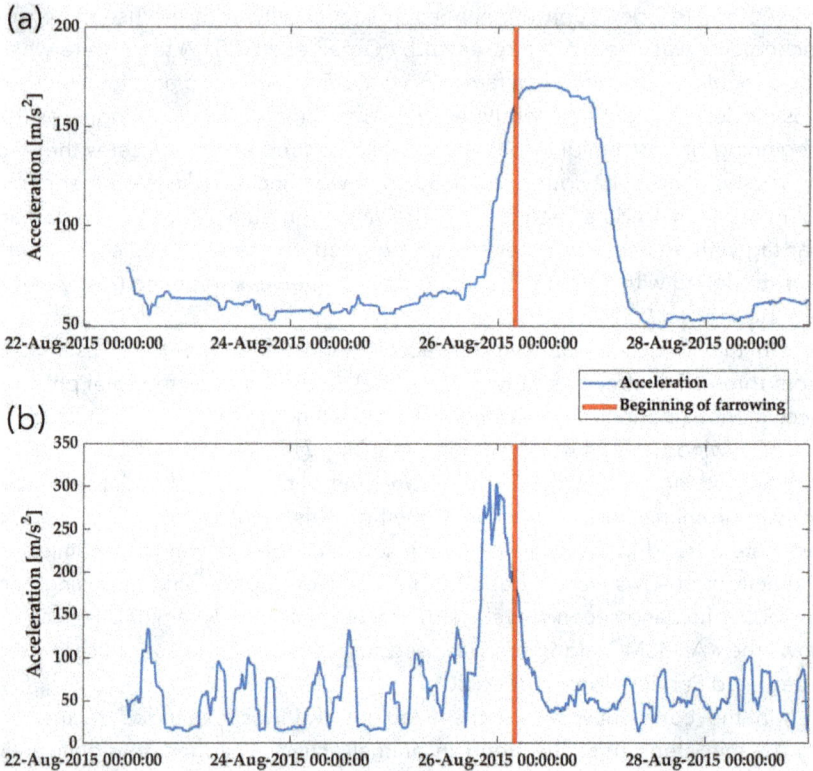

**Figure 7** Standard deviation calculated on acceleration data with two sizes of sliding windows. Period depicted in the plot starts at introduction of a sow to a farrowing pen and ends around 3 days after beginning of farrowing. (a) Sliding window of 24 h with 15 min steps. Variation in activity related to diurnal rhythms was filtered out. Two rapid changes in trend in activity were at the beginning of nest-building behaviour and at the end of nest-building behaviour, when this sow started farrowing. (b) Sliding window of 2 h with 15 min steps. Peak of activity was few hours before the onset of farrowing. From Oczak et al. 2020.

time when they are sensitive to outside disturbances (Erez and Hartsock 1990, Oliviero et al. 2008).

Austrian legislation requires that in the week before farrowing, the animals should be provided with suitable nest-building material but only when the slurry system allows it (FTT 2018). Thus, when the 'first stage' alarm is raised 'nest-building' material could be provided to the sow. Providing nest-building material in higher amounts after the 'first stage' alarm is raised or only after this alarm could limit the risk of blocking the slurry system with nest-building material and improve the welfare of sows. Additionally, the procedure of preparing a farrowing pen for newborn piglets could be started when the 'first stage' alarm is raised (Traulsen et al. 2018).

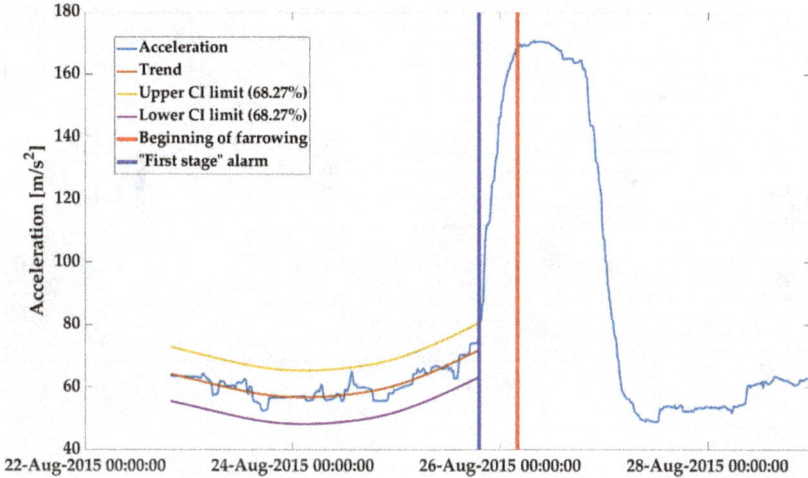

**Figure 8** 'First stage' alarm at 9 h before the onset of farrowing. Period depicted in the plot starts 12 h after introduction of a sow to a farrowing pen and ends around 3 days after farrowing. From Oczak et al. 2020.

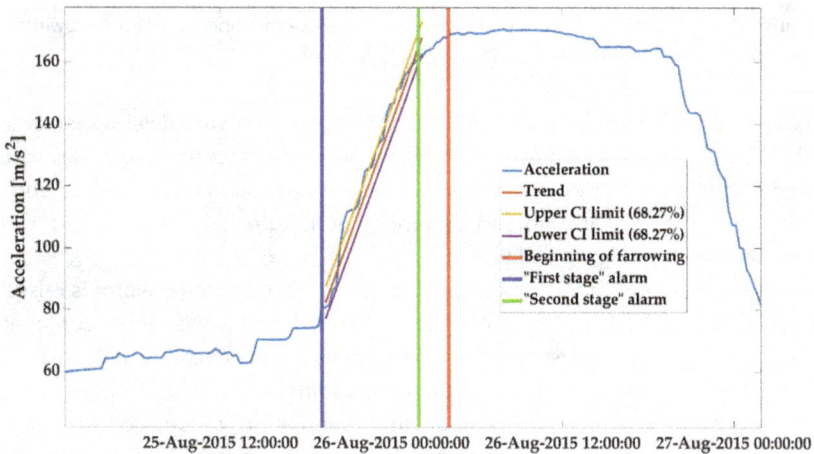

**Figure 9** 'Second stage' alarm at 2 h before the onset of farrowing. Period depicted in the plot starts around 24 h before the onset of farrowing and ends around 20 h after it. From Oczak et al. 2020.

According to Austrian Animal Welfare Directive, that will become mandatory in 2033, sows should only stay confined in farrowing crates during the 'critical period' of piglets' life (BMGÖ 2012). In order to reduce piglet-crushing, farmers will be permitted to confine sows in farrowing crates from the onset of farrowing to a few days after farrowing. From an animal welfare point of view, crating should start after nest-building is finished and before farrowing

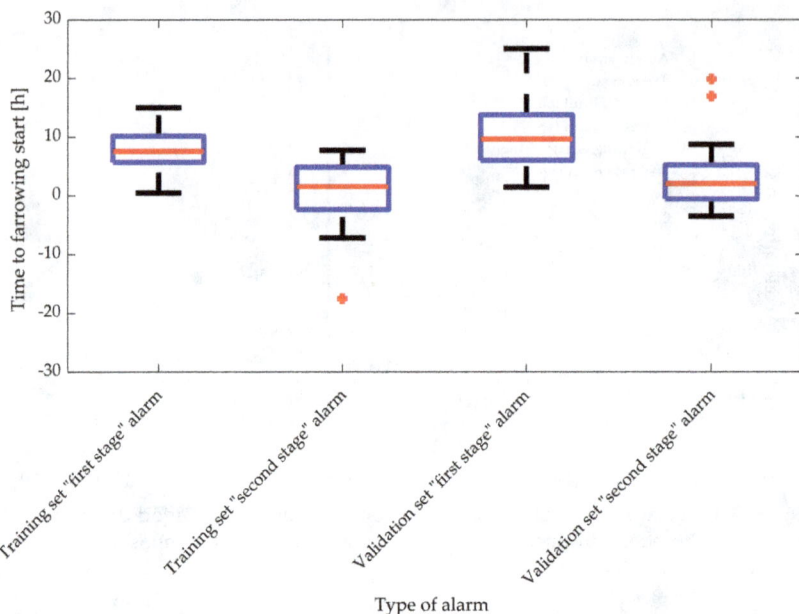

**Figure 10** Distribution of duration between time of alarms and onset of farrowing in training and validation datasets. From Oczak et al. 2020.

starts. The crate has to be opened four days after farrowing (Heidinger et al. 2017). The method used in this study to generate 'second stage' alarms is based on change of trend in sow activity when nest-building behaviour ends. This change of trend is visible as 'flattening' of feature variable extracted from acceleration data (Figs 8 and 9).

The advantage of this method is that when 'second stage' alarm is raised, most of nest-building activity of a sow should be finished. Thus, confining sows after the 'second stage' alarm was raised should create little risk for those staying in crates at the time of nest-building and especially dung times of peak of nest-building behaviour. Implementation of developed model in a monitoring system of pens with the possibility of temporary crating and practical application of the tool on farms could help to achieve a compromise between the needs of the farmer, the sow and her piglets. Especially the principle of appropriate behaviour is relevant for this research as nest-building is an essential part of behavioural repertoire in sows (WelfareQuality® 2009).

## 9 Conclusion

With all the advantages and promises of application of PLF technology for improving possibilities of nest-building behaviour in pigs there are further scientific questions which should be answered before such system could be successfully implemented in practice. For instance, can such a system attain the expected outcome considering that many first and second stage alarms would be generated in the night when mostly there would be no staff present on farms? Assuming that practical application of developed model would be to confine sows in crates just after the 'second stage' alarm is raised, many sows that farrow in the night would stay unrestrained during farrowing.

This leads to the question of whether the application of a monitoring tool, which can be used effectively only in part of farrowing animals, is justified under practical conditions. We would hypothesize that most farmers would keep the sows confined in crates over night after they've received a 'first stage' alarm, in many cases in time of nest-building behaviour, as when 'second stage' alarm is raised during the night they would not be able to intervene in the farrowing compartment. This suggests that automated confinement of sows, without the need of human intervention is an interesting topic for further research (Fig. 11). A number of scientific questions arise from an idea of automatic confinement. Is sow in the right posture that allows automated confinement and what should that posture be? In which part of the pen a sow should be located so that a crate could be automatically closed? How a sow would react when a crate would be automatically closed?

Integration of the system for the prediction of farrowing into management procedures on pig farms should also be a topic of research (Fig. 11). It is only possible to do research on this topic when models for farrowing prediction are implemented and work in real time, at least on research farms, better than on commercial farms. Further research on changes of welfare status of sows

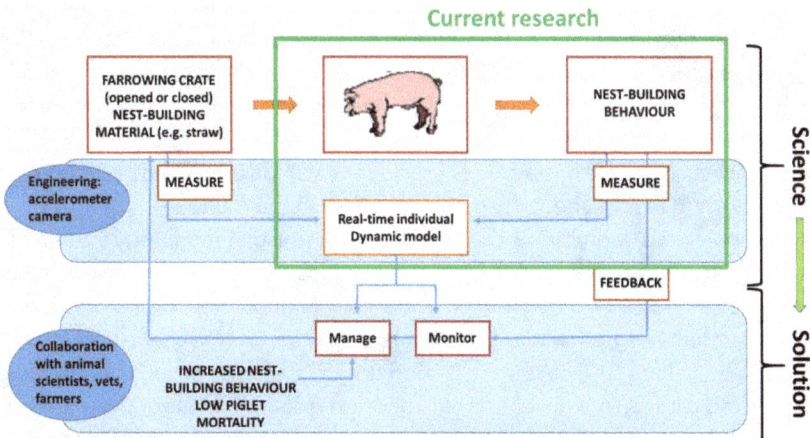

**Figure 11** How farrowing prediction in pens with possibility of temporary crating can go from science to solution. Adapted from Norton et al. 2019 and Aerts, Norton and Berckmans 2019.

on farms where such system is implemented and integrated with management procedures is necessary.

Another interesting scientific objective could be to automatically detect and differentiate between behaviours that constitute nest-building behaviour, that is, rooting, pawing, manipulation of pen or crate (Oczak et al. 2015). Such a capability could provide more detailed information on sows' behaviour and possibly increase the performance of developed models. This is more likely to be successful with image analysis techniques, which have been applied for the detection of complex behaviours in livestock but not for farrowing prediction (Viazzi et al. 2014, Lee et al. 2016).

## 10 Future trends in research

Future development of technologies for monitoring pig welfare will be dominated by three major trends: solution-oriented research, integration of key welfare indicators and wider adoption of image analysis. These major trends will be supplemented by more validation studies on open datasets and advances in robotics.

The most important trend will be solution-oriented research, where the development of an algorithm for measuring individual welfare indicator is just a first step in developing a monitoring system that could be applied in practice (Fig. 11). In order to advance from monitoring indicators of pig welfare to the development of monitoring systems that could be integrated into farm management, it is necessary to implement developed algorithms in real-time. Real-time implementation is fundamental as farm management requires immediate action (Berckmans 2013). Once real-time implementation of an algorithm is achieved, research on the use of information provided by algorithms becomes possible. Interaction between the user, the animal and technology will be at the centre of future research on monitoring welfare and management of farms with PLF tools. To provide information to various stakeholders (e.g. farm staff, vets, retailers etc.) development of real-time dashboards in PLF will become more important in the future.

Research on monitoring pig welfare with sensor technology so far has been predominantly focussed on individual indicators such as specific diseases (Ferrari et al. 2008) or behaviours (Zonderland et al. 2003). The second future trend in monitoring pig welfare will be research on integration of monitoring individual welfare indicators into monitoring systems based on multiple welfare indicators. Thus, scientists will focus on research on algorithms that will classify the overall health of an animal instead of detecting single diseases or algorithms that will evaluate if a sow performs appropriate behaviour instead of only detecting one category of behaviour (e.g. nest-building).

Image analysis has been applied for monitoring animals in farm conditions since at least 20 years (Bloemen et al. 1997). The main advantage of this sensor technology in comparison to other sensors is that it allows monitoring in a non-invasive way, without human presence or with limited human presence. It is possible to monitor multiple animals with one sensor and image-based monitoring is most similar to the way that farmers monitor their livestock with their own eyes. Camera sensor and image analysis is also probably the most versatile monitoring technology applied so far in PLF. Thus, meaning that multiple variables (key indicators) related to welfare can be monitored with one camera sensor.

However, because of difficulties in practical application in farm conditions image analysis has been used only in a limited number of products for monitoring key indicators of welfare in pigs (e.g. OptiSCAN, WUGGL One). Difficulties of practical application in pig compartments are mainly associated to the variability of environments in which pigs are kept such as changing light conditions, ceiling heights, pen sizes and also ubiquitous dust.

Nevertheless recent advantages in camera technology and image analysis techniques led to very promising developments such as detection of multiple parts of pigs body (Psota et al. 2019), face recognition (Hansen et al. 2018) or real-time pig tracking (Labrecque et al. 2019).

Further advancement in face recognition technology might have implications for intensive livestock practices globally, allowing identification of animals without the need for RFID tags, for the purposes of welfare and growth monitoring (Hansen et al. 2018). Recent research on 'Grimace Scale' in pigs (Di Giminiani et al. 2016, Viscardi et al. 2017) opens possibilities for automated estimation of affective states in pigs, including pain, with image analysis techniques.

Advances in image analysis clearly show the potential for its wider adoption in the future for monitoring pig welfare. What might remain a challenge is the lack of public datasets with images of pigs with well-defined targets and evaluation metrics (Psota et al. 2019), such as Tencent (Shenzen, China) open dataset for general-purpose image classification with over 18 million labelled images (Tencent 2018). Most scientists work with relatively small closed datasets. We foresee that this will change in the future leading to faster progress in exciting field of image analysis for welfare monitoring.

Robots are extensively used in many agricultural sectors such as crops or vegetables. So far there has been little research done on on-farm application of robots in pig industry and especially in the context of welfare monitoring. One such example includes the development of Swine Robotics (Leola, US) which offers an entire line of robots designed to improve animal welfare, safety and production. Two examples of robots developed by Swine

Robotics are 'Boar Bot' and 'Bumper Bot'. 'Function of Boar Bot' is to control the movement of a boar when heat is checked by farm staff, which should improve he welfare of the boar. Function of 'Bumper Bot' is to help in the movement of pigs for reduction of injuries in pigs. Considering that within Europe and other developed agricultural economies, the supply of stockmen with the necessary management skills is limited (Wathes et al. 2008) and that in many fields robots are more precise, consistent and faster than human workers, we envision new developments in robotics for monitoring pig welfare on farms.

## 11 Where to look for further information

The following articles provide a good overview of advances in technologies for monitoring pig welfare:

- Berckmans, D. 2006. 'Automatic on-line monitoring of animals by precision livestock farming' in Geers, R. and Madec, F., eds., Livestock Production and Society. Wageningen Academic Publishers, Wageningen, 287–295.
- Berckmans, D. 2013. 'Basic principles of PLF: gold standard, labelling and field data.' In ECPLF2015, 21–29. Leuven.
- Berckmans, D. 2017. 'General introduction to precision livestock farming', Animal Frontiers, 7: 6–11.
- Norton, T., Chen, C., Larsen, M. L. V. and Berckmans, D. 2019. 'Precision livestock farming: building 'digital representations' to bring the animals closer to the farmer', Animal: 1–9.
- Wathes, C. M., Kristensen, H. H., Aerts, J. M. and Berckmans, D. 2008. 'Is precision livestock farming an engineer's daydream or nightmare, an animal's friend or foe, and a farmer's panacea or pitfall?', Computers and Electronics in Agriculture, 64: 2–10.

Key research in this area can be found at the following organizations, journals and conferences:

- European Association for Precision Livestock Farming (EA-PLF) (www.eaplf .eu).
- Biosystems Engineering (www.journals.elsevier.com/biosystems-eng ineering).
- Computers and Electronics in Agriculture (www.journals.elsevier.com/ computers-and-electronics-in-agriculture).
- European Conference on Precision Livestock Farming (ECPLF).
- Asian Conference on Precision Livestock Farming (PLF Asia).

Major international research projects related to advances in technologies for monitoring pig welfare:

- BioBusiness (www.bio-business.eu).
- EU-PLF (www.eu-plf.eu).
- ALL-SMART PIGS (www.cordis.europa.eu/project/rcn/104741/factsheet/en).
- GroupHouseNet (www.grouphousenet.eu).

Top five research centres that readers can investigate:

- Measure, Model & Manage Bioresponses (M3-BIORES) (www.biw.kuleuve n.be/biosyst/a2h/m3-biores).
- Precision Livestock Farming Hub (PLF-Hub) (www.vetmeduni.ac.at/PLF -Hub).
- Wageningen Livestock Research (www.wur.nl/en/Research-Results/Rese arch-Institutes/livestock-research.htm).
- Teagasc Animal & Grassland Research and Innovation Centre (www .teagasc.ie).
- Research Institute for Agriculture, Fisheries and Food (ILVO) (www.ilvo.v laanderen.be/language/en-US/EN/Home.aspx).

# 12 References

Adrion, F., Kapun, A., Eckert, F., Holland, E.-M., Staiger, M., Götz, S. and Gallmann, E. (2018). Monitoring trough visits of growing-finishing pigs with UHF-RFID, *Computers and Electronics in Agriculture* 144, 144–153.

Aerts, J. M., Norton, T. and Berckmans, D. (2019). Integration of Bioresponses in management of biological processes. Course in 1st year of Masterprogramme Biosystems Engineer, Katholieke Universiteit Leuven, pp. 360, started in 2006.

Akaike, H. (1974). A new look at the statistical model identification, *IEEE Transactions on Automatic Control* 19(6), 716–723.

Ala-Kurikka, E., Heinonen, M., Mustonen, K., Peltoniemi, O., Raekallio, M., Vainio, O. and Valros, A. (2017). Behavior changes associated with lameness in sows, *Applied Animal Behaviour Science* 193, 15–20.

Algers, B. (1994). Health, behaviour and welfare of outdoor pigs, *Pig News and Information* 15(4), 113N–115N.

Alsahaf, A., Azzopardi, G., Ducro, B., Hanenberg, E., Veerkamp, R. F. and Petkov, N. (2019). Estimation of muscle scores of live pigs using a Kinect camera, *IEEE Access* 7, 52238–52245.

Amezcua, R., Walsh, S., Luimes, P. H. and Friendship, R. M. (2014). Infrared thermography to evaluate lameness in pregnant sows, *The Canadian Veterinary Journal* 55(3), 268–272.

Andersen, H. M.-L., Dybkjær, L. and Herskin, M. S. (2014). Growing pigs' drinking behaviour: number of visits, duration, water intake and diurnal variation, *Animal: An International Journal of Animal Bioscience* 8(11), 1881–1888.

Andersen, H. M.-L., Jørgensen, E., Dybkjær, L. and Jørgensen, B. (2008). The ear skin temperature as an indicator of the thermal comfort of pigs, *Applied Animal Behaviour Science* 113(1–3), 43–56.

Andretta, I., Hauschild, L., Kipper, M., Pires, P. G. S. and Pomar, C. (2018). Environmental impacts of precision feeding programs applied in pig production, *Animal* 12(9), 1990–1998.

Andretta, I., Pomar, C., Rivest, J., Pomar, J., Lovatto, P. A. and Radünz Neto, J. (2014). The impact of feeding growing-finishing pigs with daily tailored diets using precision feeding techniques on animal performance, nutrient utilization, and body and carcass composition, *Journal of Animal Science* 92(9), 3925–3936.

Andretta, I., Pomar, C., Rivest, J., Pomar, J. and Radünz, J. (2016). Precision feeding can significantly reduce lysine intake and nitrogen excretion without compromising the performance of growing pigs, *Animal* 10(7), 1137–1147.

Banhazi, T. M., Lehr, H., Black, J., Crabtree, H., Schofield, P., Tscharke, M. and Berckmans, D. (2012). Precision livestock farming: an international review of scientific and commercial aspects, *International Journal of Agricultural and Biological Engineering* 5(3), 1–9.

Berckmans, D. (2006). Automatic on-line monitoring of animals by precision livestock farming. In: Geers, R. and Madec, F. (Eds), *Livestock Production and Society*. Wageningen: Wageningen Academic Publishers, pp. 287–295.

Berckmans, D. (2013). Basic principles of PLF: gold standard, labelling and field data. In: Berckmans, D. and Vandermeulen, J. (Eds), *6th European Conference on Precision Livestock Farming*, Leuven, Belgium, 10–12 September 2013. The Organising Committee of the 6th European Conference on Precision Livestock Farming (ECPLF), pp. 21–29.

Berckmans, D. (2014). Precision livestock farming technologies for welfare management in intensive livestock systems, *Revue Scientifique et Technique* 33(1), 189–196.

Berckmans, D. (2017). General introduction to precision livestock farming, *Animal Frontiers: The Review Magazine of Animal Agriculture* 7(1), 6–11.

Bloemen, H., Aerts, J.-M., Berckmans, D. and Goedseels, V. (1997). Image analysis to measure activity index of animals, *Equine Veterinary Journal* 29(Suppl. 23), 16–19.

Blömke, L., Volkmann, N. and Kemper, N. (2020). Evaluation of an automated assessment system for ear and tail lesions as animal welfare indicators in pigs at slaughter, *Meat Science* 159, 107934.

BMGÖ (2012). Verordnung des Bundesministers für Gesundheit, mit der die 1. Tierhaltungsverordnung geändert wird, *Bundesgesetzblatt für die Republik Österreich*, BGBl. II Nr. 61/2012.

Boissy, A., Manteuffel, G., Jensen, M. B., Moe, R. O., Spruijt, B., Keeling, L. J., Winckler, C., Forkman, B., Dimitrov, I., Langbein, J., Bakken, M., Veissier, I. and Aubert, A. (2007). Assessment of positive emotions in animals to improve their welfare, *Physiology and Behavior* 92(3), 375–397.

Bos, J. M., Bovenkerk, B., Feindt, P. H. and Van Dam, Y. K. (2018). The quantified animal: precision livestock farming and the ethical implications of objectification, *Food Ethics* 2(1), 77–92.

Boumans, I. J. M. M., de Boer, I. J. M., Hofstede, G. J. and Bokkers, E. A. M. (2018). How social factors and behavioural strategies affect feeding and social interaction patterns in pigs, *Physiology and Behavior* 194, 23–40.

Brown-Brandl, T., Maselyne, J., Adrion, F., Kapun, A., Hessel, E., Saeys, W., Van Nuffel, A. and Gallmann, E. (2017). Comparing three different passive RFID systems for behaviour monitoring in grow-finish pigs. In: *Proceedings of the* 8th European Conference on *Precision Livestock Farming*, Nantes, France, 12–14 September 2017, pp. 622–631.

Brünger, J., Dippel, S., Koch, R. and Veit, C. (2019). 'Tailception': using neural networks for assessing tail lesions on pictures of pig carcasses, *Animal* 13(5), 1030–1036.

Castrén, H., Algers, B., De Passille, A.-M., Rushen, J. and Uvnäs-Moberg, K. (1993). Preparturient variation in progesterone, prolactin, oxytocin and somatostatin in relation to nest building in sows, *Applied Animal Behaviour Science* 38(2), 91–102.

Chen, C., Zhu, W., Guo, Y., Ma, C., Huang, W. and Ruan, C. (2018). A kinetic energy model based on machine vision for recognition of aggressive behaviours among group-housed pigs, *Livestock Science* 218, 70–78.

Chen, C., Zhu, W., Liu, D., Steibel, J., Siegford, J., Wurtz, K., Han, J. and Norton, T. (2019). Detection of aggressive behaviours in pigs using a RealSence depth sensor, *Computers and Electronics in Agriculture* 166, 105003.

Chen, C., Zhu, W., Steibel, J., Siegford, J., Wurtz, K., Han, J. and Norton, T. (2020). Recognition of aggressive episodes of pigs based on convolutional neural network and long short-term memory, *Computers and Electronics in Agriculture* 169, 105166.

Chung, Y., Oh, S., Lee, J., Park, D., Chang, H.-H. and Kim, S. (2013). Automatic detection and recognition of pig wasting diseases using sound data in audio surveillance systems, *Sensors* 13(10), 12929–12942.

Condotta, I. C. F. S., Brown-Brandl, T. M., Silva-Miranda, K. O. and Stinn, J. P. (2018). Evaluation of a depth sensor for mass estimation of growing and finishing pigs, *Biosystems Engineering* 173, 11–18.

Conte, S., Bergeron, R., Gonyou, H., Brown, J., Rioja-Lang, F. C., Connor, L. and Devillers, N. (2014). Measure and characterization of lameness in gestating sows using force plate, kinematic, and accelerometer methods, *Journal of Animal Science* 92(12), 5693–5703.

Conte, S., Bergeron, R., Gonyou, H., Brown, J., Rioja-Lang, F. C., Connor, M. L. and Devillers, N. (2015). Use of an analgesic to identify pain-related indicators of lameness in sows, *Livestock Science* 180, 203–208.

Cook, N. J., Chabot, B., Lui, T., Bench, C. J. and Schaefer, A. L. (2015). Infrared thermography detects febrile and behavioural responses to vaccination of weaned piglets, *Animal* 9(2), 339–346.

Cornou, C. and Lundbye-Christensen, S. (2008). Classifying sows' activity types from acceleration patterns, *Applied Animal Behaviour Science* 111(3–4), 262–273.

Cornou, C., Vinther, J. and Kristensen, A. R. (2008). Automatic detection of oestrus and health disorders using data from electronic sow feeders, *Livestock Science* 118(3), 262–271.

Costa, A., Ismayilova, G., Borgonovo, F., Viazzi, S., Berckmans, D. and Guarino, M. (2014). Image-processing technique to measure pig activity in response to climatic variation in a pig barn, *Animal Production Science* 54(8), 1075–1083.

D'Eath, R. B., Jack, M., Futro, A., Talbot, D., Zhu, Q., Barclay, D. and Baxter, E. M. (2018). Automatic early warning of tail biting in pigs: 3D cameras can detect lowered tail posture before an outbreak, *PLoS ONE* 13(4), e0194524.

da Silva, J. P., de Alencar Nääs, I., Abe, J. M. and da Silva Cordeiro, A. F. (2019). Classification of piglet (*Sus scrofa*) stress conditions using vocalization pattern and applying paraconsistent logic Eτ, *Computers and Electronics in Agriculture* 166, 105020.

Di Giminiani, P., Brierley, V. L., Scollo, A., Gottardo, F., Malcolm, E. M., Edwards, S. A. and Leach, M. C. (2016). The assessment of facial expressions in piglets undergoing tail docking and castration: toward the development of the piglet grimace scale, *Frontiers in Veterinary Science* 3, 100.

Diana, A., Carpentier, L., Piette, D., Boyle, L. A., Berckmans, D. and Norton, T. (2019). An ethogram of biter and bitten pigs during an ear biting event: first step in the development of a Precision Livestock Farming tool, *Applied Animal Behaviour Science* 215, 26-36.

Erez, B. and Hartsock, T. G. (1990). A microcomputer-photocell system to monitor periparturient activity of sows and transfer data to remote location, *Journal of Animal Science* 68(1), 88-94.

Escalante, H. J., Rodriguez, S. V., Cordero, J., Kristensen, A. R. and Cornou, C. (2013). Sow-activity classification from acceleration patterns: a machine learning approach, *Computers and Electronics in Agriculture* 93, 17-26.

EU-PLF (2016). Project. Final Report. Bright Farm by Precision Livestock Farming. Grant agreement no: 311825.

European Food Safety Authority (2007). The risks associated with tail biting in pigs and possible means to reduce the need for tail docking considering the different housing and husbandry systems – Scientific Opinion of the Panel on Animal Health and Welfare, *EFSA Journal* 5(12), 611.

European Union (2008a). Council Directive 2008/71/EC of 15 July 2008 on the identification and registration of pigs, *Official Journal of the European Union* L 213, 31-36.

European Union (2008b). Council Directive 2008/120/EC of 18 December 2008 laying down minimum standards for the protection of pigs, *Official Journal of the European Union* L 47, 5-13

European Union (2017). Regulation (EU) 2017/745 of the European Parliament and of the Council of 5 April 2017 on medical devices, amending Directive 2001/83/EC, Regulation (EC) No 178/2002 and Regulation (EC) No 1223/2009 and repealing Council Directives 90/385/EEC and 93/42/EEC, *Official Journal of the European Union* L 117, 1-175.

Exadaktylos, V., Silva, M., Aerts, J.-M., Taylor, C. J. and Berckmans, D. (2008). Real-time recognition of sick pig cough sounds, *Computers and Electronics in Agriculture* 63(2), 207-214.

Faucitano, L. (2001). Causes of skin damage to pig carcasses, *Canadian Journal of Animal Science* 81(1), 39-45.

Fels, M., Konen, K., Hessel, E. and Kemper, N. (2018). Determination of static space occupied by individual weaner and growing pigs using an image-based monitoring system, *Journal of Agricultural Science* 156(2), 282-290.

Fernández, A. P., Norton, T., Youssef, A., Exadaktylos, V., Bahr, C., Bruininx, E., Vranken, E. and Berckmans, D. (2019). Real-time modelling of individual weight response to feed supply for fattening pigs, *Computers and Electronics in Agriculture* 162, 895-906.

Ferrari, S., Silva, M., Guarino, M. and Berckmans, D. (2008). Analysis of cough sounds for diagnosis of respiratory infections in intensive pig farming, *Transactions of the ASABE* 51(3), 1051-1055.

FTT (2018). *Handbuch Schweine*, Fachstelle für tiergerechte Tierhaltung und Tierschutz. Available at: https://www.tierschutzkonform.at/wp-content/uploads/tierschutzkon form.at-handbuch-schweine-2auflage2018.pdf.

Gregoire, J., Bergeron, R., D'Allaire, S., Meunier-Salaun, M. C. and Devillers, N. (2013). Assessment of lameness in sows using gait, footprints, postural behaviour and foot lesion analysis, *Animal* 7(7), 1163-1173.

Hansen, M. F., Smith, M. L., Smith, L. N., Salter, M. G., Baxter, E. M., Farish, M. and Grieve, B. (2018). Towards on-farm pig face recognition using convolutional neural networks, *Computers in Industry* 98, 145-152.

Heidinger, B., Stinglmayr, J. and Baumgartner, J. (2017). Projekt Pro-SAU: Ergebnisse zur Evaluierung von neuen Abferkelbuchten mit Bewegungsmöglichkeit für die Sau, *Leipziger Blaue Hefte* 469-470.

Heinonen, M., Peltoniemi, O. and Valros, A. (2013). Impact of lameness and claw lesions in sows on welfare, health and production, *Livestock Science* 156(1-3), 2-9.

Held, S. D. E. and Špinka, M. (2011). Animal play and animal welfare, *Animal Behaviour* 81(5), 891-899.

Hunter, E. J., Jones, T. A., Guise, H. J., Penny, R. H. C. and Hoste, S. (2001). The relationship between tail biting in pigs, docking procedure and other management practices, *The Veterinary Journal* 161(1), 72-79.

Iida, R., Piñeiro, C., Koketsu, Y. (2017). Behavior, displacement and pregnancy loss in pigs under an electronic sow feeder, *Journal of Agricultural Science* 9(12), 43-53.

Jarvis, S., D'Eath, R. B., Robson, S. K. and Lawrence, A. B. (2006). The effect of confinement during lactation on the hypothalamic-pituitary–adrenal axis and behaviour of primiparous sows, *Physiology and Behavior* 87(2), 345-352.

Jorgenson, D. W. and Vu, K. M. (2016). The ICT revolution, world economic growth, and policy issues, *Telecommunications Policy* 40(5), 383-397.

Jun, K., Kim, S. J. and Ji, H. W. (2018). Estimating pig weights from images without constraint on posture and illumination, *Computers and Electronics in Agriculture* 153, 169-176.

Kammersgaard, T. S., Malmkvist, J. and Pedersen, L. J. (2013). Infrared thermography – a non-invasive tool to evaluate thermal status of neonatal pigs based on surface temperature, *Animal* 7(12), 2026-2034.

Karriker, L. A., Abell, C. E., Pairis-Garcia, M. D., Holt, W. A., Sun, G., Coetzee, J. F., Johnson, A. K., Hoff, S. J. and Stalder, K. J. (2013). Validation of a lameness model in sows using physiological and mechanical measurements, *Journal of Animal Science* 91(1), 130-136.

Kashiha, M., Bahr, C., Haredasht, S. A., Ott, S., Moons, C. P. H., Niewold, T. A., Ödberg, F. O. and Berckmans, D. (2013). The automatic monitoring of pigs water use by cameras, *Computers and Electronics in Agriculture* 90, 164-169.

Kashiha, M., Bahr, C., Ott, S., Moons, C. P. H., Niewold, T. A., Ödberg, F. O. and Berckmans, D. (2014). Automatic weight estimation of individual pigs using image analysis, *Computers and Electronics in Agriculture* 107, 38-44.

King, R. L., Baxter, E. M., Matheson, S. M. and Edwards, S. A. (2019). Temporary crate opening procedure affects immediate post-opening piglet mortality and sow behaviour, *Animal* 13(1), 189-197.

Kongsro, J. (2014). Estimation of pig weight using a Microsoft Kinect prototype imaging system, *Computers and Electronics in Agriculture* 109, 32-35.

Kuzuhara, Y., Kawamura, K., Yoshitoshi, R., Tamaki, T., Sugai, S., Ikegami, M., Kurokawa, Y., Obitsu, T., Okita, M., Sugino, T. and Yasuda, T. (2015). A preliminarily study for predicting body weight and milk properties in lactating Holstein cows using a three-dimensional camera system, *Computers and Electronics in Agriculture* 111, 186-193.

Labrecque, J., Gouineau, F. and Rivest, J. (2019). Real-time individual pig tracking and behavioural metrics collection with affordable security cameras. In: O'Brien, B., Hennessy, D. and Shalloo, L. (Eds), *9th European Conference on Precision Livestock Farming*, Cork, Ireland, 26-29 August 2019. The Organising Committee of the 9th European Conference on Precision Livestock Farming (ECPLF), Teagasc, Animal & Grassland Research and Innovation Centre, Moorepark, Fermoy, Co. Cork, pp. 460-466.

Lahrmann, H. P., Hansen, C. F., D'Eath, R., Busch, M. E. and Forkman, B. (2018). Tail posture predicts tail biting outbreaks at pen level in weaner pigs, *Applied Animal Behaviour Science* 200, 29-35.

Lambertz, C., Petig, M., Elkmann, A. and Gauly, M. (2015). Confinement of sows for different periods during lactation: effects on behaviour and lesions of sows and performance of piglets, *Animal* 9(8), 1373-1378.

Larsen, M. L. V., Andersen, H. M.-L. and Pedersen, L. J. (2019a). Changes in activity and object manipulation before tail damage in finisher pigs as an early detector of tail biting, *Animal* 13(5), 1037-1044.

Larsen, M. L. V., Pedersen, L. J. and Jensen, D. B. (2019b). Prediction of tail biting events in finisher pigs from automatically recorded sensor data, *Animals* 9(7), 458.

Lawrence, A. B., Newberry, R. C. and Špinka, M. (2018) 15 - .Positive welfare: what does it add to the debate over pig welfare? In: Špinka, M. (Ed.), *Advances in Pig Welfare*. Duxford: Woodhead Publishing, 415-444.

Lee, J., Jin, L., Park, D. and Chung, Y. (2016). Automatic recognition of aggressive behavior in pigs using a Kinect depth sensor, *Sensors* 16(5), 631.

Leidig, M. S., Hertrampf, B., Failing, K., Schumann, A. and Reiner, G. (2009). Pain and discomfort in male piglets during surgical castration with and without local anaesthesia as determined by vocalisation and defence behaviour, *Applied Animal Behaviour Science* 116(2-4), 174-178.

Lu, M., He, J., Chen, C., Okinda, C., Shen, M., Liu, L., Yao, W., Norton, T. and Berckmans, D. (2018). An automatic ear base temperature extraction method for top view piglet thermal image, *Computers and Electronics in Agriculture* 155, 339-347.

Madsen, T. N., Andersen, S. and Kristensen, A. R. (2005). Modelling the drinking patterns of young pigs using a state space model, *Computers and Electronics in Agriculture* 48(1), 39-61.

Madsen, T. N. and Kristensen, A. R. (2005). A model for monitoring the condition of young pigs by their drinking behaviour, *Computers and Electronics in Agriculture* 48(2), 138-154.

Maes, D. G. D., Janssens, G. P. J., Delputte, P., Lammertyn, A. and de Kruif, A. (2004). Back fat measurements in sows from three commercial pig herds: relationship with reproductive efficiency and correlation with visual body condition scores, *Livestock Production Science* 91(1-2), 57-67.

Manteuffel, G., Puppe, B. and Schön, P. C. (2004). Vocalization of farm animals as a measure of welfare, *Applied Animal Behaviour Science* 88(1-2), 163-182.

Marcet Rius, M., Pageat, P., Bienboire-Frosini, C., Teruel, E., Monneret, P., Leclercq, J., Lafont-Lecuelle, C. and Cozzi, A. (2018). Tail and ear movements as possible indicators of emotions in pigs, *Applied Animal Behaviour Science* 205, 14-18.

Marchant, J., Rudd, A., Mendl, M., Broom, D., Meredith, M., Corning, S. and Simmins, P. (2000). Timing and causes of piglet mortality in alternative and conventional farrowing systems, *Veterinary Record* 147, 214.

Martínez-Avilés, M., Fernández-Carrión, E., López García-Baones, J. M. and Sánchez-Vizcaíno, J. M. (2015). Early detection of infection in pigs through an online monitoring system, *Transboundary and Emerging Diseases* 64(2), 364-373.

Maselyne, J., Adriaens, I., Huybrechts, T., De Ketelaere, B., Millet, S., Vangeyte, J., Van Nuffel, A. and Saeys, W. (2016a). Measuring the drinking behaviour of individual pigs housed in group using radio frequency identification (RFID), *Animal* 10(9), 1557-1566.

Maselyne, J., Saeys, W., Briene, P., Mertens, K., Vangeyte, J., De Ketelaere, B., Hessel, E. F., Sonck, B. and Van Nuffel, A. (2016b). Methods to construct feeding visits from RFID registrations of growing-finishing pigs at the feed trough, *Computers and Electronics in Agriculture* 128, 9-19.

Maselyne, J., Saeys, W., De Ketelaere, B., Mertens, K., Vangeyte, J., Hessel, E. F., Millet, S. and Van Nuffel, A. (2014). Validation of a High Frequency Radio Frequency Identification (HF RFID) system for registering feeding patterns of growing-finishing pigs, *Computers and Electronics in Agriculture* 102, 10-18.

Maselyne, J., Van Nuffel, A., Briene, P., Vangeyte, J., De Ketelaere, B., Millet, S., Van den Hof, J., Maes, D. and Saeys, W. (2017). Online warning systems for individual fattening pigs based on their feeding pattern, *Biosystems Engineering* 173, 143-156.

Matthews, S. G., Miller, A. L., Plötz, T. and Kyriazakis, I. (2017). Automated tracking to measure behavioural changes in pigs for health and welfare monitoring, *Scientific Reports* 7(1), 1-12.

McGlone, J. J. (1985). A quantitative ethogram of aggressive and submissive behaviors in recently regrouped pigs, *Journal of Animal Science* 61(3), 556-566.

McGlone, J. J., Kelley, K. W. and Gaskins, C. T. (1980). Lithium and porcine aggression, *Journal of Animal Science* 51(2), 447-455.

McKenna, S., Amaral, T., Plötz, T. and Kyriazakis, I. (2018). Multi-part segmentation for porcine offal inspection with auto-context and adaptive atlases, *Pattern Recognition Letters* 112, 290-296.

McManus, C., Tanure, C. B., Peripolli, V., Seixas, L., Fischer, V., Gabbi, A. M., Menegassi, S. R. O., Stumpf, M. T., Kolling, G. J., Dias, E. and Costa, J. B. G. (2016). Infrared thermography in animal production: an overview, *Computers and Electronics in Agriculture* 123, 10-16.

Meijer, E., Oosterlinck, M., van Nes, A., Back, W. and van der Staay, F. J. (2014). Pressure mat analysis of naturally occurring lameness in young pigs after weaning, *BMC Veterinary Research* 10(1), 193.

Melišová, M., Illmann, G., Andersen, I. L., Vasdal, G. and Haman, J. (2011). Can sow pre-lying communication or good piglet condition prevent piglets from getting crushed?, *Applied Animal Behaviour Science* 134(3-4), 121-129.

Mohling, C. M., Johnson, A. K., Coetzee, J. F., Karriker, L. A., Abell, C. E., Millman, S. T. and Stalder, K. J. (2014). Kinematics as objective tools to evaluate lameness phases in multiparous sows, *Livestock Science* 165, 120-128.

Moinard, C., Mendl, M., Nicol, C. J. and Green, L. E. (2003). A case control study of on-farm risk factors for tail biting in pigs, *Applied Animal Behaviour Science* 81(4), 333-355.

Moura, D. J., Silva, W. T., Naas, I. A., Tolón, Y. A., Lima, K. A. O. and Vale, M. M. (2008). Real time computer stress monitoring of piglets using vocalization analysis, *Computers and Electronics in Agriculture* 64(1), 11-18.

Moustsen, V. A., Hales, J. and Hansen, C. F. (2012). Farrowing systems with temporary crating. Report of the Free Farrowing Workshop Vienna 2011.

Munsterhjelm, C., Heinonen, M. and Valros, A. (2015). Effects of clinical lameness and tail biting lesions on voluntary feed intake in growing pigs, *Livestock Science* 181(Suppl. C), 210-219.

Nasirahmadi, A., Hensel, O., Edwards, S. A. and Sturm, B. (2016). Automatic detection of mounting behaviours among pigs using image analysis, *Computers and Electronics in Agriculture* 124(Suppl. C), 295-302.

Nasirahmadi, A., Hensel, O., Edwards, S. A. and Sturm, B. (2017). A new approach for categorizing pig lying behaviour based on a Delaunay triangulation method, *Animal* 11(1), 131-139.

Norton, T., Chen, C., Larsen, M. L. V. and Berckmans, D. (2019). Review: Precision livestock farming: building 'digital representations' to bring the animals closer to the farmer, *Animal: An International Journal of Animal Bioscience* 13(12), 3009-3017.

Oczak, M. (2018). Precision Livestock Farming for sows and weaner pigs. PhD thesis, Catholic University of Leuven.

Oczak, M., Maschat, K. and Baumgartner, J. (2019). Monitoring of approaching farrowing in pens with possibility of temporary crating on the basis of ear tag acceleration data. In: O'Brien, B., Hennessy, D. and Shalloo, L. (Eds), *European Conference on Precision Livestock Farming 2019*, Cork, Ireland, 26-29 August 2019. The Organising Committee of the 9th European Conference on Precision Livestock Farming (ECPLF), Teagasc, Animal & Grassland Research and Innovation Centre, Moorepark, Fermoy, Co. Cork, pp. 322-329.

Oczak, M., Maschat, K. and Baumgartner, J. (2020). Dynamics of sows' activity housed in farrowing pens with possibility of temporary crating might indicate the time when sows should be confined in a crate before the onset of farrowing, *Animals* 10(1), 6.

Oczak, M., Maschat, K., Berckmans, D., Vranken, E. and Baumgartner, J. (2015). Classification of nest-building behaviour in non-crated farrowing sows on the basis of accelerometer data, *Biosystems Engineering* 140(Suppl. C), 48-58.

Oczak, M., Viazzi, S., Ismayilova, G., Sonoda, L. T., Roulston, N., Fels, M., Bahr, C., Hartung, J., Guarino, M., Berckmans, D. and Vranken, E. (2014). Classification of aggressive behaviour in pigs by activity index and multilayer feed forward neural network, *Biosystems Engineering* 119, 89-97.

Oliviero, C., Pastell, M., Heinonen, M., Heikkonen, J., Valros, A., Ahokas, J., Vainio, O. and Peltoniemi, O. A. T. (2008). Using movement sensors to detect the onset of farrowing, *Biosystems Engineering* 100(2), 281-285.

Pedersen, L. J., Malmkvist, J. and Andersen, H. M.-L. (2013). Housing of sows during farrowing: a review on pen design, welfare and productivity. In: Aland, A. and Banhazi, T. (Eds), *Livestock Housing: Modern Management to Ensure Optimal Health and Welfare of Farm Animals*. Wageningen: Wageningen Academic Publishers, pp. 285-297.

Pezzuolo, A., Guarino, M., Sartori, L., González, L. A. and Marinello, F. (2018). On-barn pig weight estimation based on body measurements by a Kinect v1 depth camera, *Computers and Electronics in Agriculture* 148, 29-36.

Pluym, L. M., Maes, D., Vangeyte, J., Mertens, K., Baert, J., Van Weyenberg, S., Millet, S. and Van Nuffel, A. (2013). Development of a system for automatic measurements of force and visual stance variables for objective lameness detection in sows: SowSIS, *Biosystems Engineering* 116(1), 64-74.

Pomar, C., Hauschild, L., Zhang, G.-H., Pomar, J. and Lovatto, P. A. (2009). Applying precision feeding techniques in growing-finishing pig operations, *Revista Brasileira de Zootecnia* 38(spe), 226-237.

Pomar, C. and Remus, A. (2019). Precision pig feeding: a breakthrough toward sustainability, *Animal Frontiers: The Review Magazine of Animal Agriculture* 9(2), 52-59.

Psota, E. T., Mittek, M., Pérez, L. C., Schmidt, T. and Mote, B. (2019). Multi-pig part detection and association with a fully-convolutional network, *Sensors* 19(4), 852.

Reimert, I., Bolhuis, J. E., Kemp, B. and Rodenburg, T. B. (2013). Indicators of positive and negative emotions and emotional contagion in pigs, *Physiology and Behavior* 109(Suppl. C), 42-50.

Reiners, K., Hegger, A., Hessel, E. F., Böck, S., Wendl, G. and Van den Weghe, H. F. A. (2009). Application of RFID technology using passive HF transponders for the individual identification of weaned piglets at the feed trough, *Computers and Electronics in Agriculture* 68(2), 178-184.

Royakkers, L., Timmer, J., Kool, L. and van Est, R. (2018). Societal and ethical issues of digitization, *Ethics and Information Technology* 20(2), 127-142.

Savulescu, C. (2015). Dynamics of ICT development in the EU, *Procedia Economics and Finance* 23, 513-520.

Scheel, C., Traulsen, I., Auer, W., Muller, K., Stamer, E. and Krieter, J. (2017). Detecting lameness in sows from ear tag-sampled acceleration data using wavelets, *Animal* 11(11), 2076-2083.

Schon, P., Puppe, B. and Manteuffel, G. (2004). Automated recording of stress vocalisations as a tool to document impaired welfare in pigs, *Animal Welfare-Potters Bar Then Wheathampstead* 13(2), 105-110.

Shao, B. and Xin, H. (2008). A real-time computer vision assessment and control of thermal comfort for group-housed pigs, *Computers and Electronics in Agriculture* 62(1), 15-21.

Shi, C., Teng, G. and Li, Z. (2016). An approach of pig weight estimation using binocular stereo system based on LabVIEW, *Computers and Electronics in Agriculture* 129, 37-43.

Singh, C., Verdon, M., Cronin, G. M. and Hemsworth, P. H. (2017). The behaviour and welfare of sows and piglets in farrowing crates or lactation pens, *Animal* 11(7), 1210-1221.

Soerensen, D. D. and Pedersen, L. J. (2015). Infrared skin temperature measurements for monitoring health in pigs: a review, *Acta Veterinaria Scandinavica* 57(1), 5.

Statham, P., Green, L., Bichard, M. and Mendl, M. (2009). Predicting tail-biting from behaviour of pigs prior to outbreaks, *Applied Animal Behaviour Science* 121(3-4), 157-164.

Stavrakakis, S., Li, W., Guy, J. H., Morgan, G., Ushaw, G., Johnson, G. R. and Edwards, S. A. (2015). Validity of the Microsoft Kinect sensor for assessment of normal walking patterns in pigs, *Computers and Electronics in Agriculture* 117(Suppl. C), 1-7.

Stevenson, P. (2017). Precision livestock farming: could it drive the livestock sector in the wrong direction. Compassion in World Farming. Available at: https://www.ciwf.org.uk/media/7431928/plf-could-it-drive-the-livestock-sector-in-the-wrong-direction.pdf.

Šustr, P., Špinka, M., Cloutier, S. and Newberry, R. C. (2001). Computer-aided method for calculating animal configurations during social interactions from two-dimensional coordinates of color-marked body parts, *Behavior Research Methods, Instruments, and Computers* 33(3), 364-370.

Taylor, A. A. and Weary, D. M. (2000). Vocal responses of piglets to castration: identifying procedural sources of pain, *Applied Animal Behaviour Science* 70(1), 17-26.

Tencent (2018). Tencent ML images. Available at: https://github.com/Tencent/tencent-ml -images/tree/master/data.

Thomsen, R., Bonde, M., Kongsted, A. G. and Rousing, T. (2012). Welfare of entire males and females in organic pig production when reared in single-sex groups, *Livestock Science* 149(1-2), 118-127.

Traulsen, I., Breitenberger, S., Auer, W., Stamer, E., Muller, K. and Krieter, J. (2016). Automatic detection of lameness in gestating group-housed sows using positioning and acceleration measurements, *Animal* 10(6), 970-977.

Traulsen, I., Scheel, C., Auer, W., Burfeind, O. and Krieter, J. (2018). Using acceleration data to automatically detect the onset of farrowing in sows, *Sensors* 18(1), 170.

Turner, S. P., Farnworth, M. J., White, I. M. S., Brotherstone, S., Mendl, M., Knap, P., Penny, P. and Lawrence, A. B. (2006). The accumulation of skin lesions and their use as a predictor of individual aggressiveness in pigs, *Applied Animal Behaviour Science* 96(3-4), 245-259.

Ursinus, W. W., Van Reenen, C. G., Kemp, B. and Bolhuis, J. E. (2014). Tail biting behaviour and tail damage in pigs and the relationship with general behaviour: predicting the inevitable?, *Applied Animal Behaviour Science* 156, 22-36.

Valletta, J. J., Torney, C., Kings, M., Thornton, A. and Madden, J. (2017). Applications of machine learning in animal behaviour studies, *Animal Behaviour* 124, 203-220.

van Putten, G. (1969). An investigation into tail-biting among fattening pigs, *British Veterinary Journal* 125(10), 511-517.

Verdouw, C., Wolfert, S. and Tekinerdogan, B. (2016). Internet of Things in agriculture, *CAB Reviews: Perspectives in Agriculture, Veterinary Science, Nutrition and Natural Resources* 11(35), 1-12.

Viazzi, S., Ismayilova, G., Oczak, M., Sonoda, L. T., Fels, M., Guarino, M., Vranken, E., Hartung, J., Bahr, C. and Berckmans, D. (2014). Image feature extraction for classification of aggressive interactions among pigs, *Computers and Electronics in Agriculture* 104, 57-62.

Viscardi, A. V., Hunniford, M., Lawlis, P., Leach, M. and Turner, P. V. (2017). Development of a piglet grimace scale to evaluate piglet pain using facial expressions following castration and tail docking: a pilot study, *Frontiers in Veterinary Science* 4, 51.

Vranken, E. and Berckmans, D. (2017). Precision livestock farming for pigs, *Animal Frontiers: The Review Magazine of Animal Agriculture* 7(1), 32-37.

Wallenbeck, A. and Keeling, L. J. (2013). Using data from electronic feeders on visit frequency and feed consumption to indicate tail biting outbreaks in commercial pig production, *Journal of Animal Science* 91(6), 2879-2884.

Wang, K., Guo, H., Ma, Q., Su, W., Chen, L. and Zhu, D. (2018). A portable and automatic Xtion-based measurement system for pig body size, *Computers and Electronics in Agriculture* 148, 291-298.

Wang, X., Zhao, X., He, Y. and Wang, K. (2019). Cough sound analysis to assess air quality in commercial weaner barns, *Computers and Electronics in Agriculture* 160, 8-13.

Wathes, C. M., Kristensen, H. H., Aerts, J.-M. and Berckmans, D. (2008). Is precision livestock farming an engineer's daydream or nightmare, an animal's friend or foe, and a farmer's panacea or pitfall?, *Computers and Electronics in Agriculture* 64(1), 2-10.

Weary, D. M., Braithwaite, L. A. and Fraser, D. (1998). Vocal response to pain in piglets, *Applied Animal Behaviour Science* 56(2-4), 161-172.

Weber, A., Salau, J., Haas, J. H., Junge, W., Bauer, U., Harms, J., Suhr, O., Schönrock, K., Rothfuß, H., Bieletzki, S. and Thaller, G. (2014). Estimation of backfat thickness using extracted traits from an automatic 3D optical system in lactating Holstein-Friesian cows, *Livestock Science* 165, 129–137.

Wechsler, B. (1995). Coping and coping strategies: a behavioural view, *Applied Animal Behaviour Science* 43(2), 123–134.

Wedin, M., Baxter, E. M., Jack, M., Futro, A. and D'Eath, R. B. (2018). Early indicators of tail biting outbreaks in pigs, *Applied Animal Behaviour Science* 208, 7–13.

WelfareQuality® (2009). *Welfare Quality Assessment Protocol for Pigs (Sows and Piglets, Growing and Finishing Pigs).* Lelystad: WelfareQuality® Consortium, 122.

Werkheiser, I. (2018). Precision livestock farming and farmers' duties to livestock, *Journal of Agricultural and Environmental Ethics* 31(2), 181–195.

Wischner, D., Kemper, N. and Krieter, J. (2009). Nest-building behaviour in sows and consequences for pig husbandry, *Livestock Science* 124(1–3), 1–8.

Xin, H. (1999). Assessing swine thermal comfort by image analysis of postural behaviors, *Journal of Animal Science* 77(Suppl. 2), 1–9.

Yeates, J. W. and Main, D. C. J. (2008). Assessment of positive welfare: a review, *The Veterinary Journal* 175(3), 293–300.

Young, P. C. (1984). *Recursive Estimation and Time-Series Analysis: An Introduction.* Chicago, IL: Springer Science & Business Media.

Young, P. C. (2011). *Recursive Estimation and Time-Series Analysis: An Introduction for the Student and Practitioner.* Berlin: Springer Science & Business Media.

Zonderland, J. J., Schepers, F., Bracke, M. B. M., den Hartog, L. A., Kemp, B. and Spoolder, H. A. M. (2011). Characteristics of biter and victim piglets apparent before a tail-biting outbreak, *Animal* 5(5), 767–775.

Zonderland, J. J., Vermeer, H., Vereijken, P. and Spoolder, H. (2003). Measuring a pig's preference for suspended toys by using an automated recording technique. *International Commission of Agricultural and Biosystems Engineering (CIGR) E-Journal* 5.

9 781801 460538